LUBRICANTS,

OILS AND GREASES.

LUBRICANTS, OILS AND GREASES.

TREATED THEORETICALLY AND GIVING PRACTICAL
INFORMATION REGARDING THEIR

COMPOSITION, USES AND MANUFACTURE.

*A PRACTICAL GUIDE FOR MANUFACTURERS, ENGINEERS,
AND USERS IN GENERAL OF LUBRICANTS*

BY

ILTYD I. REDWOOD,

ASSOCIATE MEMBER AMERICAN SOCIETY OF MECHANICAL ENGINEERS;
MEMBER SOCIETY CHEMICAL INDUSTRY (ENGLAND);
AUTHOR OF 'THEORETICAL AND PRACTICAL AMMONIA REFRIGERATION,' AND 'A PRACTICAL
TREATISE ON MINERAL OILS AND THEIR BY-PRODUCTS.'

New York:
SPON & CHAMBERLAIN, 12 CORTLANDT STREET
London:
E. & F. N. SPON, Limited, 125 STRAND.
1898

Entered according to Act of Congress, in the year 1897,
By ILTYD I. REDWOOD,
in the office of the Librarian of Congress, Washington, D. C.

TROW DIRECTORY
PRINTING AND BOOKBINDING COMPANY
NEW YORK

PREFACE.

IN dealing with the subject embraced by this little work, the author has attempted to give Engineers an insight to the properties of the various lubricants that are likely to be offered them, and thus enable them to guard against the choice of one that would be likely to prove unsatisfactory for the purpose for which it is intended.

Also, the author has endeavoured to explain, as concisely as possible, to those desiring information regarding the manufacture of lubricants—and greases in special—the theory and general requirements that govern their manufacture.

BANTRY HOUSE, BELVEDERE,
KENT, ENGLAND, 1897.

CONTENTS.

INTRODUCTION: LUBRICANTS, ... xi

THEORETICAL.

CHAPTER I.

MINERAL OILS: AMERICAN AND RUSSIAN; HYDROCARBONS, . 1

CHAPTER II.

FATTY OILS: GLYCERIDES, ... 6
VEGETABLE OILS; FISH OILS, ... 9

CHAPTER III.

MINERAL LUBRICANTS, ... 10

CHAPTER IV.

GREASES: COMPOUNDED; "SET" OR AXLE; "BOILED" OR CUP, 11

CHAPTER V.

TESTS OF OILS: MINERAL OILS, ... 15
TESTS OF OILS: FATTY OILS, ... 18

MANUFACTURE.

CHAPTER VI.

	PAGE
MINERAL OIL LUBRICANTS,	19
COMPOUNDED OILS,	20
DEBLOOMED OILS,	22

CHAPTER VII.

GREASES: COMPOUNDED GREASES,	23
"SET" OR AXLE GREASES,	24
BOILED GREASES,	28
ENGINE GREASES,	36
COMPOUNDED GREASES,	38

APPENDIX.

THE ACTION OF OILS ON VARIOUS METALS, ... 43

INDEX, ... 51

TABLES.

	PAGE
I. Viscosity and Specific Gravity,	2
II. Atomic Weights,	13
III. Origin, Tests, Etc., of Oils,	16
IV. Action of Oils on Metals,	45

LIST OF PLATES.

I. I. I. Redwood's Improved Set Measuring Apparatus.

II. Section Grease Kettle.

III. Diagram of Action of Oils on Metals.

INTRODUCTION.

LUBRICANTS.

In order to better enable Engineers to choose a lubricant suitable for their special purpose, it will undoubtedly be advisable to explain the chemical composition of oils and greases, the characteristics of the various fatty and mineral oils that are met with, and the changes in their composition that take place under certain conditions.

Lubricants embrace a variety of products which may be divided into the two classes:

Oils.
Greases.

The former includes mineral, fatty, and other oils, while the latter is represented by all compounds of oils and fats that are solid or semi-solid at the ordinary temperature.

THEORETICAL.

CHAPTER I.

MINERAL OILS.

OILS under this heading are hydrocarbons, which are derived from the distillation of the products of the American and Russian oil wells and bituminous shales, etc.

The hydrocarbons represented in the above oils belong to the Paraffin and Olefine series, and are generally spoken of as paraffin, or mineral lubricating oils.

Paraffins and olefines differ in that a paraffin contains two atoms more hydrogen than the corresponding member of the olefine group; for example, the lowest member of each group is, respectively:

Marsh gas...................... CH_4
Olefiant gas.................... CH_2

American and Russian lubricating mineral oils contain a larger percentage of paraffins than olefines, while the shale mineral oils usually are the reverse, and contain more olefines than paraffins. This difference between natural and shale oils is undoubtedly due to the fact that the latter oils have to be distilled more often than the natural—Russian or American—oils in order to bring them to the same state of refinement, and as the distillation of a paraffin oil results in the decomposition of certain members of the paraffin group, with a deposition of carbon and formation of an olefine, it follows that the more often a hydrocarbon is distilled the more olefines it will contain. As there is but little doubt that paraffins are better lubricants than olefines, the fact that natural oils yield better lubricants than can be obtained from shale oils, is easily accounted for.

Mineral lubricating oils are not affected by high-pressure steam or alkalies, and these characteristics enable them to be used where other lubricants would be quite unfitted for the work.

The tests that usually influence the choice or purchase of lubricating oils are the specific gravity and viscosity; but in reality these tests count for little, for the reason that but a small idea of the lubricating quality of an oil can be gained from either of the tests separately, and not very much even when taken conjointly, except in extreme cases; as, for instance, we know that a thin light oil is useless for steam cylinder or heavy journal lubrication, and that a thick heavy oil is equally useless for light high-speed spindles.

To show how little can be judged of the consistency of an oil from the specific gravity, it is only necessary to refer to Table I.

TABLE I.

Specific Gravity.			Viscosity.
875	Shale Oil		121
875	American Ordinary Oil		63
875	" Spindle "		214
920	" Ordinary "		405
920	Russian " "		560

On the other hand, the viscosity is no criterion of the lubricating power, because a high viscosity oil may owe a large percentage of its viscosity to the presence of wax, and in that case its lubricating efficiency will be low as compared with another oil of the same viscosity but of a lower cold test (*i.e.*, containing less wax).

It is now seen that the lubricating value of an oil cannot be accurately judged from either the specific gravity or the viscosity; and also, if a certain oil—910, for example—acts as an efficient lubricant for a certain piece of machinery it does not prove that an oil of the same specific gravity but another "make" will answer the purpose equally well, and neither does it follow that an oil of lower specific gravity will not do the work.

Some readers will naturally say, "Why not have mechanical appliances by means of which the lubricating quality can be determined?" Such appliances have been in use a number of years by various lubricant manufacturers, but from the experience the author has had with these "testing machines" he considers the results obtained by their use of very little practical value, for the reason that the journals, etc., of these "testing machines" are as perfect as care and machinery can make them, while the journals, etc., of ordinary machinery are usually less perfect, and, consequently, an oil that acts admirably in a testing machine is generally a failure when used on the machinery for which it is intended.

Mineral oils have certain properties that render them invaluable for use in special cases where either a pure fatty or a compounded oil would be more or less injurious. For instance, in oiling the stuffing-boxes of the ammonia cylinders of freezing machines nothing but a pure, high-class mineral oil should be used, because the action of ammonia on fatty oils results in the formation of a soap, and, as more or less oil always escapes past the packing and finds its way into the cylinder, it is obvious that if a fatty lubricant were used it would be converted into a soap and would not be long in finding its way into the condenser coils, etc., and give endless trouble.

In the case of steam cylinders—and especially where high-pressure steam is used—a fatty oil should not be allowed to enter into the composition of the lubricant, for the reason, as stated under Fatty Oils, that when a fatty oil is brought in contact with high-pressure steam a decomposition takes place, and the resulting products exert a considerable destructive action on the piston and cylinder.

On the other hand, a mineral oil of low specific gravity and flash-point must not be used, as its lubricating power will be entirely inadequate. Nothing but a high-grade cylinder oil should be used for steam-cylinder lubrication, and the flash-point should not be lower than 400° F.

Also, an oil finished with an acid and soda treatment is not fitted for steam-cylinder lubrication, as such oil always contains more or less sulphur compounds which are driven off by the heat, and will eventually cause trouble.

High-grade cylinder oils (which are comparatively dark in colour) are reduced or concentrated oils which have either been well settled or else filtered through animal charcoal.

MINERAL OILS.

There are a large number of so-called *debloomed* oils on the market, and it is of the utmost importance that engineers should guard against their use. These oils are made for the sole purpose of enabling unprincipled lubricant manufacturers (especially middlemen) to sophisticate fatty oils with a cheaper mineral oil. As the deblooming of a mineral oil can only be accomplished by treating the oil with nitric acid or a nitro-compound such as nitro-naphthaline, nitro-toluole, etc., it stands to reason that the lubrication of machinery by a debloomed oil means ruination to the machinery. The only really sure way of engineers avoiding the probability of debloomed oils being present in their lubricants is to buy from reputable firms and have nothing to do with middlemen who handle debloomed oils.

It is of great importance to remember that the higher the temperature the less is the lubricating power of any lubricant, and consequently, if we have an efficient lubricant under normal conditions, it may be totally useless for the purpose of reducing the friction of a bearing that has become suddenly heated, due to any of the causes that lead to such "engineers' annoyances." If possible, do not allow a bearing to become heated—by giving it proper attention; but if it does, do not waste time and material by applying its lubricant, but rather go to work and remove the cause of the trouble, and allow the bearing to cool before trying to start again.

CHAPTER II.

FATTY OILS.

FATTY oils differ from hydrocarbons in that they contain oxygen as well as hydrogen and carbon. Those fatty oils that concern engineers may be said, in a general way, to consist of

$$\left.\begin{array}{l}\text{Oleic}\\ \text{Palmitic}\\ \text{Stearic}\end{array}\right\} \text{glycerides.}$$

These glycerides are very susceptible to decomposition, or turn what is commonly called rancid. A fresh oil or fat is neutral, that is, neither acid nor alkaline, but a rancid fat gives an acid reaction, and in that state is very deleterious to the metals usually employed in the construction of machinery. The decomposition of an oil or fat takes place more rapidly in a warm, moist atmosphere than in a moderately cool, dry one, and for that reason it is obvious that if a fatty oil is brought in contact with steam its decomposition will soon be effected. The discovery of this fact and its practical application revolutionised the manufacture of oleic, palmitic, and stearic acids, which, in former days, were obtained by decomposing the fats, etc., by saponification and then liberating the fatty acids by neutralising the alkali with

FATTY OILS.

an acid; while nowadays the same result is obtained by subjecting the fats, etc., to the action of steam which has been heated to about 500° F. The reaction that takes place is as follows:

Supposing we take a fat—stearine, for example—and subject it to the action of the above-heated steam, three atoms of carbon and five of hydrogen disassociate themselves with the stearine and combine with the oxygen and part of the hydrogen of the steam and thus form glycerine, while the rest of the carbon and hydrogen, and also the oxygen atoms of the stearine, combine with the rest of the hydrogen of the steam and form stearic acid. This reaction is shortly explained by the following equation:

$$\underbrace{C_3H_5(C_{18}H_{35}O)_3}_{\text{Stearine.}} + \underbrace{3H_2O}_{\substack{\text{Water in form} \\ \text{of steam.}}} = \underbrace{3C_{18}H_{36}O_2}_{\text{Stearic acid.}} + \underbrace{C_3H_5(HO)_3}_{\text{Glycerine.}}$$

Hot stearic, palmitic, or oleic acids readily attack copper and its alloys as well as steel, iron, etc., and therefore it is not difficult to see from the above how it is that steam cylinders and pistons are eaten away when a fatty oil, either by itself or in admixture with mineral oils, is used for lubricating them. So great is the action that the author has removed from a 10″ x 12″ cylinder two pounds of a very hard black substance (that had been rolled into balls by the action of the piston) that contained

    31 per cent. of fatty matter,
    23 per cent. of moisture,
    46 per cent. of iron,

and which was the result of using a fatty oil lubricant.

The piston and cylinder showed unmistakable proof of the action of the oil, although the latter had only been in use some six or eight months, and the pressure of steam used in this case had not exceeded forty pounds per square inch.

Instances where fatty oils alone can be advantageously used are very few, but one case is in the use of taps and dies. These should never be lubricated with anything but pure lard or neatsfoot oils. It is, of course, apparently cheaper to use a mineral oil or some of the so-called "screw-cutting" compounded oils, but their use invariably results in the ruination of the taps and dies in a fraction of the time that they would last if properly lubricated.

Mixtures of fatty and mineral oils, or, as they are called, compounded oils, have innumerable uses; but it is impossible to lay down any fixed law that would enable an engineer to choose the right compound for any special machinery, for the reason that no two bearings are absolutely alike or will work equally well with the same oil; but there is, of course, the one general rule governing the choice of lubricants, namely : — Light machinery that runs at a very high speed—such as loom spindles, etc.—should be lubricated with a limpid, free-flowing, low specific gravity oil; while heavy machinery of low speed needs a thick-bodied lubricant; and heavy machinery of high speed needs a moderately thick-bodied lubricant.

The only sure way of doing justice to machinery is to try numerous lubricants, and when one is obtained that

FATTY OILS.

gives thoroughly satisfactory results, use it and no other —even though an offer is made of a "far superior article at a less price."

It seems hardly necessary to state that neither raw nor boiled linseed oils must be used as a lubricant, as they quickly oxidise and "gum."

Castor and cotton-seed oils are often used in compounds, but their use should not be encouraged, as they are far more liable to gum, and therefore retard rather than aid lubrication, than:

> Colza oil,
> Rape-seed oil,
> Olive oil,
> Palm oil,

which latter, together with the animal and fish oils,

> Neatsfoot,
> Lard,
> Tallow,
> Sperm,
> Whale,

are the best fatty oils for use in compounding. Menhaden fish oil should not be used in lubricants, as it soon becomes rancid and gums.

CHAPTER III.

MINERAL LUBRICANTS.

THERE are a number of compounds on the market sold under various names, which are more or less transparent and of a jelly-like or gelatinous consistency.

These mineral lubricants are nothing more or less than either stearates, palmitates, or oleates of alumina mixed with a certain proportion (according to the consistency) of mineral oil. Such lubricants are bad for the reasons that—

1st. They have a great tendency to separate into a thick jelly and free mineral oil, the former of which will choke the oil-holes and prevent lubrication.

2d. Owing to the uncertainty of the feed the bearings are liable to become overheated.

3d. Heated bearings will liquefy the jelly, and in that condition it has little or no lubricating quality.

On the other hand, there are anti-friction metals (which are often and successfully used) which might be classed under this heading, because, as they require no lubricants (in the general meaning of the word), they are lubricants in themselves.

Graphite or plumbago and soapstone are also often used very successfully without any admixture, for the lubrication of excessively heavy and slow-running machinery.

CHAPTER IV.

GREASES.

Good greases are undoubtedly the best and most economical lubricants for machinery if properly applied. They may be divided into three classes, viz:—
Compounded,
"Set," or Axle.,
"Boiled," or "Cup,"

Compounded greases represent a very undesirable class, as they are simply mechanical mixtures of oils and fats, or fats and waxes, or all three. If kept for any length of time they oxidise and become very rancid, and in that state are quite unfit for contact with metals. If they contain paraffin, bees' or other waxes, they might be better called friction producers than lubricants. Compounded greases though cheap of themselves are in reality most expensive, as they are usually inefficient and destructive.

Set greases represent a class of cheap greases that are made from mineral and rosin oils, which are converted into greases by the agency of lime. These greases, as their *alias* implies, are largely used for lubricating the axles of carts, etc., but are more suitable for greasing the inside curves of coal runs, street railroads, etc., and for protecting wire cables, etc., from the weather.

Boiled greases belong to the class that calls for special attention, because it provides the most theoretically and practically perfect lubricant for medium and heavy machinery.

A good boiled grease is perfectly neutral and will remain so for an indefinite length of time. It is made by converting fats or fatty oils into an insoluble* (in water) soap and dissolving the soap in mineral oil.

Boiled greases are usually made from horse fat or cotton-seed oil, but other fats or oils can be used. The fats or oils used in making these greases contain:—

 (1) Stearine—$C_3H_5(C_{18}H_{35}O_2)_3$;
 (2) Palmatine—$C_3H_5(C_{16}H_{31}O_2)_3$;
 (3) Oleine—$C_3H_5(C_{18}H_{33}O_2)_3$;

and we will suppose that the average composition is:—

 (4) $C_3H_5(C_{17}H_{33}O_2)_3$.

If two equivalents of (4) are boiled with 3 equivalents of slaked lime, six equivalents of stearate-palmitic-oleate of lime (soap) and two equivalents of glycerine will be obtained, thus:—

$$\underbrace{2C_3H_5(C_{17}H_{33}O_2)_3}_{\text{Fat and Oil.}} + \underbrace{3Ca(HO)_2}_{\text{Slaked Lime.}} =$$

$$\underbrace{6C_{17}H_{34}O_2Ca}_{\text{Soap.}} + \underbrace{2C_3H_5(HO)_3}_{\text{Glycerine.}}$$

In order to determine what the equivalent weights are, we must refer to the atomic weights given in Table II.

---

* Soaps made by the use of soda or potash are soluble, but if made by the use of lime they are insoluble.

TABLE II.

Element.	Symbol.	Atomic Weight.
Hydrogen =	H =	1
Oxygen =	O =	16
Carbon =	C =	12
Calcium =	Ca =	40
Sodium =	Na =	23
Potassium =	K =	39
Aluminium =	Al =	27.5

From the above we figure that:—$C_3H_5(C_{17}H_{33}O_2)_3$, or $C_{54}H_{104}O_6$, weighs 848 parts, and as we have to take 2 equivalents we have 1,696 parts. $Ca(HO)_2$ weighs 74, or the 3 equivalents = 222 parts. It is therefore apparent that in order to convert (theoretically) 1,696 parts by weight of the above fats into an *insoluble* soap, it will be necessary to use 222 parts by weight of slaked lime.

As the atomic *values* of soda or potash are only half that of calcium, only one equivalent of fat would be used for 3 equivalents of NaHO or KHO in the event of a *soluble* soap being required.

In concluding the theoretical portion of this subject considerable stress should be laid on the fact that the value or even purity of a mineral oil must not be judged from appearances. A dark-coloured oil is often purer than a lighter coloured one, and in oiling ordinary machinery the consumer would act wisely if he used oils that had been finished with a distillation in preference to those that had been finished with a "treatment." The former may range from a yellowish brown to dark brown in colour by transmitted light, but providing they have been either settled or filtered so as to free them from any fine particles of coke, etc., they are far better lubricants than the same oils after acid and soda treatments.

Colour counts for nothing in the lubricating quality of an oil, and the only case where good colour is of any importance is when the oil is used for the lubrication of spindles, etc., of looms and is liable to drip on, or otherwise come in contact with, the material that is being manufactured.

CHAPTER V.

TESTS OF OILS.

THE tests of the various oils and fats used in the manufacture of liquid and solid lubricants are given in Table III., a careful study of which, together with an experience in the smell, taste, etc., of the products mentioned will be of great value to both lubricant manufacturers and engineers.

There are numerous tests which are credited with enabling chemists to distinguish between fatty oils of various origin, but, whereas these tests are more or less reliable in the hands of experienced operators for determining the origin of any single oil, they are practically worthless when a complex mixture has to be dealt with, and therefore these tests will not be further noticed. However, it is often important to know whether a certain oil is either a pure mineral or a pure fatty oil, and methods for determining these points will now be considered.

MINERAL OILS.

It is important that these should be free from sulphur compounds, and whether this is so or not may be judged by heating a small quantity to a temperature of 300° F. and maintaining that temperature for about fif-

TABLE III.

Name.	Type.	Origin.	Specific Gravity.	Melting Point.	Viscosity.	Consistency at 70° F.	Colour.
Beeswax	Insect	Bees	963	150		Solid	
Castor oil	Vegetable	Ricinus communis	969			Fluid	
Colza	"	Brassica campestris oleifera	914			"	
Cotton-seed	"	Gossypium barbadense	931			"	
Cylinder "A"	Mineral	American crude	894		146+	Semi-fluid	Brown black
" "Cold Test"	"	"	886		116+	semi-solid	Reddish green
" "Continental"	"	"	886		105+	"	Brownish green
" "Extra Filtered"	"	"	886		159+	"	Reddish green
" "F. F. F."	"	"	889		151+	Fluid	Reddish
" "Green"	"	"	894		166+	Semi-fluid	Greenish brown
" "Locomotive"	"	"	897		199+	"	Brown black
" "N"	"	"	897		170+	"	"
Degras	Animal	Wool washing	996			Solid	Brown black
Horse fat	"	Horses	918			"	Brown
Japan wax	Vegetable	Rhus succedanea, vernicifera and sylvestris		130		"	Light yellow
Lard oil	Animal	Pig	915			Fluid	
Menhaden fish	Fish	Alosa menhaden (species of herring)	931			"	
Neatsfoot	Animal	Oxen's feet	913			"	
Olive oil	Vegetable	Olea europœa	916			"	
Oleine	" and animal	Olive oil and animal fats		90		Semi-solid	Reddish
Oleic acid	" "	"	897			Fluid	
Pale oil	Mineral	American crude	865		50		
"	"	"	876		68		
"	"	"	885		103		
"	"	"	910		928		
"	"	"	915		275		
"	"	"	920		405		
"	"	"	920		560		
"	"	"	875		121		
"	"	Russian	968				
Palm oil	"	Scotch shale		95		Solid	Yellow
Palmatine	Vegetable	Elœis guineensis					
Palmytic acid	" and animal	Palm oil and animal fats				Semi-fluid	Brown black
Pine tar	Vegetable	"				Fluid	Yellow
Rape-seed oil (blown)	"	"	1,068			"	"
" " (summer)	"	Brassica præcox	916			"	
" " (winter)	"	Brassica napus	917			"	
Rosin oil, 1st run	"	"	1,009				
" 2d "	"	"	995				
" 3d "	"	"	956			Solid	
Sperm oil	Fish	Physeter macrocephalus (sperm whale)	876				
Stearic acid	Animal	Animal fats		130			
Stearine	"	"		101			
Tallow	"	"		102			
Valve oil, xxx dark	Mineral	American crude	915			Liquid	
" " light	"	"	887		159+	Semi-solid	
Whale oil	Fish	Balena mysticetus (Greenland whale)	887		149+	Fluid	
West Virginia	Mineral	American crude	923		70	"	Brownish

+ = at 212° F — other viscosities at 70° F.

teen minutes, then allowing to cool, and comparing the colour with that of the original oil. If any considerable darkening has taken place it is proof that the oil has been badly refined and is unfit for use in steam cylinders or on bearings that are liable to "heat."

In order to determine whether the oil has been artificially thickened by means of an aluminium or other metallic soap, take a small (20 or 25 grains) weighed quantity and ignite in a crucible, and finally subject to the full heat of a Bunsen burner until all the carbon has been burned off. If any ash remains, weigh it, and if it is over 0.1 per cent. its existence is doubtless due to the presence of a metallic soap in the oil. The nature of the metal in the ash is of course easily determined by tests known to all chemists.

If the presence of a fatty oil is suspected, boil a small quantity of the oil in a test-tube with about 10 per cent. of its bulk alcoholic soda (made by dissolving caustic soda in alcohol), and if, on cooling, the oil becomes somewhat gelatinous or even becomes solid it is proof that a fatty oil is present. In order to determine the percentage of fatty oil present it will be necessary to take about 25 to 50 grains of the oil, the same weight of fine silver sand, and about 12 grains of alcoholic soda and evaporate in a porcelain dish, with constant stirring, over a water-bath. After all traces of alcohol have been driven off, the residue should be repeatedly washed with gasoline and the washings passed through a filter, which treatment will result in the extraction by the gasoline of all the mineral oil, and if the filtrates are very cautiously evaporated from a tared flask until all trace of gasoline has disappeared the residue can be weighed and will consist of the mineral oil contained in the original oil, and its percentage deducted from the original will give the percentage of fatty oil.

Fatty Oils.

The presence of mineral oils in fatty oils is easily detected by placing a drop of the suspected oil on a glass plate (the back of which has been painted with a mixture of shellac varnish and lampblack) and holding the latter at various angles to the light; if mineral oil is present the drop of oil will show a distinct "bloom" or florescence.

The presence of a debloomed mineral oil is not quite so easily detected as an undebloomed, but, if debloomed mineral has been added to the fatty oil in any considerable quantity its presence can be detected by the "black plate;" and small quantities can be detected by heating the oil to 300° or 400° F., when disagreeable smelling fumes of nitrous acid will be given off.

The quantity of mineral oil that may be present in a fatty oil can be determined in the manner described for determining the percentage of fatty oil in a mineral oil.

A useful practical test to apply to a lubricant is to take a bright steel needle and immerse it in the lubricant, which should then be maintained at a temperature of about 100° F. for two or three days. If after that time the needle shows the slightest indication of rusting, the use of the lubricant should be discontinued at once.

MANUFACTURE.

CHAPTER VI.

MINERAL OIL LUBRICANTS.

THE manufacture of mineral lubricating oils will not be particularly dealt with here as the author has treated the subject fully in another work,* but certain differences in the manufacture will here be noted, as they are of importance to lubricant manufacturers and users.

These oils are dividable into the two classes:—

Distillates,

Reduced (or, more properly, "concentrated") oils.

The former constitute the oils manufactured from shale products and certain American and Russian oils. As their name implies, they are oils that have been distilled off from the contents of the still. Distillates are treated in various ways to bring their flash-point up to a certain required point, and are then purified by acid and soda treatments.

Reduced or concentrated oils are, on the other hand, made by charging the still and, instead of running it to

* "Mineral Oils and their By-Products." Same publishers as this book.

dryness, carefully firing it, and, when its contents have been brought to the boiling-point, reducing the fire to as moderate a heat as possible while a large volume of steam (that is superheated to the temperature of the oil) is forced into the oil through a perforated pipe at a point near the bottom of the still, thus effecting the concentration of the oil without subjecting it to undue external and "rash" heat. By this method of working a concentrated oil of exceptionally high viscosity is obtained, and can be drawn off from the still after the latter has been allowed to cool for a few hours. The still must, of course, be fitted with a draw-off and sampling cocks, and as soon as the tests of a sample show that the desired viscosity and specific gravity is reached the fire is drawn and steam shut off. When concentrated oils are properly manufactured they not only have a good colour and far higher viscosity, but also have far greater lubricating powers than a distillate of the same specific gravity; and if oil refiners in general adopted the proper method of concentrating their lubricating oils, instead of manufacturing distillates it would undoubtedly lead to an increased trade in this class of lubricants, and a decrease in the use of many of the noxious compounds.

Compounded Oils.

These are innumerable and variable in their composition. Amongst other things that often enter into their composition may be mentioned :—

India-rubber,
Rosin,

Aluminium soaps,
Lead soaps,
Blown rape-seed oil.

The addition of any of the above to a mineral oil is made with a view to increasing the viscosity or body, it being cheaper to obtain the desired viscosity in this way than to buy a mineral oil that naturally has that desired viscosity. The people who make these compounded oils are sometimes ignorant and sometimes aware of the fact that the above additions ultimately result in the gumming of the bearings and serious loss of power—not to speak of undue wear and tear. If an engineer complains to one of these sellers that his oils gum and the seller knows his business, and the engineer is not up in oils, the former simply tells the latter that he should pass some kerosene oil through his bearings now and again, and the trouble will cease. This advice is, of course, correct, but if the seller supplied a lubricant of first-rate quality there would not be any need for the engineer to resort to the use of kerosene at all.

Compounded oils include :—

Engine oils,	Harness oils,
Cylinder oils,	Leather oils,
Spindle oils,	Rope oils,
Cordage oils,	Screw-cutting oils,
Dynamo oils,	Polishing oils,
Electric light oils,	Wool oils,

and many others, the composition of which will not be given here, as this is not a book of receipts.

Debloomed Oils.

As this class of oils is prepared solely with a view to attempting to defraud the users of fatty oils, the manufacture will receive no further notice than was given it in the "Theoretical" chapter.

CHAPTER VII.

GREASES.

In order to deal with the manufacture of greases, it will be necessary to divide them into three classes, viz.:

>Compounded,
>"Set," or axle,
>"Boiled," or cup.

COMPOUNDED GREASES.

These greases require and deserve but a few words of explanation, as they are simply mechanical mixtures of animal fats, paraffin wax, or vegetable fats with any suitable mineral oil. When paraffin wax enters into the combination the grease is indeed a poor one and unfit for use on valuable machinery. When animal or vegetable fats are contained in these kinds of greases the lubricating power is greatly superior to that of paraffin wax greases, but as the fats are more or less easily oxidisable, the greases usually soon become rancid and unfit for use.

Compounded greases include non-lubricants, such as:

>Belt grease,
>Harness grease,
>Leather grease,
>Polishing grease, etc.

"SET," OR AXLE GREASES

Include a large number of greases that are sold under various trade names, but are all manufactured on the same principle, although the constituents are various. The manufacture is effected as follows:

To a mixture of a mineral oil and slaked lime—called the "lime part"—is added a certain quantity of rosin oil—called the "set"—and the whole having been well mixed, will, in a short space of time, set and become solid or semi-solid, according to the percentage of "set" used.

The proportions of lime and mineral oil in the lime part vary somewhat with the quality of the slaked lime. If the manufacturer wishes to slake his own lime he must be careful in the choice of the stone, and a stone that slakes slowly and quietly is preferable to one that slakes quickly and with a great evolution of heat. Of the ordinary limestones the "Jointa" is best suited for the purpose, but a marble lime is preferable to any other. The slaking should be done in a brick-lined shallow pit, and the least possible quantity of water should be used, and after the whole of the stone is thoroughly slaked it should be turned over two or three times a day until perfectly dry, and then it must be passed through a hundred-mesh brass wire sieve or bolting machine, and after that it is most important that it should be kept entirely free from contact with air until used. If sifted slaked lime has free contact with air it will absorb carbonic acid and thus lose a large percentage of its causticity, and be of far less value than a lime from which the air has been excluded.

Few manufacturers will find it remunerative to slake their own lime, as there are now so many dealers in slaked lime who cater to the necessities of the "set grease" manufacturers.

Whether the slaked lime be home-made or bought, it will be necessary to make small experiments to determine the minimum quantity that is necessary to give the hardest grease with the least quantity of rosin oil. As a general thing, about one pound of slaked lime should be mixed with every two gallons of mineral oil, and the percentage of rosin oil that is added to this mixture may vary from two to twenty-five, according to the hardness of the grease required.

On the manufacturing scale, say, 1,000 gallons of mineral oil (of the desired grade), and 500 pounds of sifted slaked lime are placed in a tank fitted with a mechanical stirrer and thoroughly agitated for eight or ten hours, after which the contents of the tank are ready for use, but the agitation must be continued while the tank is being drawn from. As this stock of lime part may be sufficient to last for a number of days, it must be remembered that a certain quantity of the lime will settle in the tank, and therefore the agitation must be started one-half to two hours (according to the length of time the contents of tank have been at rest) before any of this "stock" is used; if this precaution is not strictly followed it will be found that a considerable increase in the percentage of rosin oil will be required to give a grease of the same consistency as would be obtained if the settled lime had been given ample time to become thoroughly incorporated with the oil again.

If the grease is to be sold in barrels, the requisite quantity of the lime part is run into the barrel, and by being vigorously stirred, by quickly raising and lowering a perforated metal disc (about seven inches in diameter) attached to an iron rod or handle, the rosin oil or "set" is quickly run in (see Plate I.) and the whole stirred for a few seconds and then allowed to remain at perfect rest until thoroughly set.

When the grease is to be filled into small packages, such as one-pound boxes, etc., not more than two gallons should be mixed at a time. The lime part should be run into a spouted pail of about two and one-half gallons capacity, and while the set is being quickly added the contents of the pail should receive three or four vigorous stirs with the hand and arm, and then be immediately and quickly poured into the small packages. If the contents of the pail commence to set before being emptied into the packages, it is necessary that what is left in the pail should be scraped out into a barrel and afterwards either mixed in with the contents of the lime part stock-tank or passed between rollers, or through a paint mill, or otherwise worked until its granular consistency has been reduced to a smooth homogeneous grease.

In order to ensure a set grease being perfectly smooth and free from "grain" or lumps, it is necessary that the following particulars should be carefully noted :—

1st. Both the lime part and set must be as cold as possible (*i.e.*, so that they will flow freely and are not too thick to mix easily), say 70° F. for the former and 75° to 80° F. for the latter.

2d. The floor on which the packages are filled must be entirely free from any vibration, and no attempt must be made to remove the packages after their contents have once commenced to set until the grease is thoroughly set. The length of time required for setting varies according to the quantity of set used, temperature of the air, etc., but usually ranges from 5 to 15 minutes.

3d. If an excess of water is present in either the lime part or set it is liable to separate from the grease or make the latter lumpy. Separation of water is especially liable to take place if the grease ingredients are too warm. It might be stated, in a general way, that if a thin mineral oil is used it must not contain more than three per cent. of water, while the use of a thick mineral oil will often permit of the presence of seven per cent. of water.

Set greases include:

> Black grease,
> Cake or cold-roll grease,
> Curve grease,
> Ferry-slip grease,
> Gear grease,
> Graphite grease,
> Hub axle grease.
> Mica grease,
> Pinion grease,
>   etc., etc.

Boiled Greases.

This class is dividable into two varieties known as:—
"Cup" greases,
"Engine" greases.

The former greases are undoubtedly the best lubricants that can be used, and they may be manufactured from either horse fat or from cotton or rape-seed oils. The manufacture of this variety of grease consists in first converting the fat or oil into an insoluble soap, and afterwards dissolving this soap in a mineral oil.

The saponification, or conversion of the fat or oil into a soap, is accomplished by the use of milk of lime, with the aid of heat and constant agitation. Formerly, the "kettles" in which these greases were made were heated by means of a fire; but this method of heating was unsatisfactory in that it was dangerous, did not allow of proper and necessary fine regulation, often resulted in a large waste of materials, due to boiling over, and at the best did not give uniform results or grade of grease. The use of steam-jacketted vessels not only does away with the danger from fire and the probability of a boil over, but also results in the production of a far better grease than can be obtained by furnace heat.

The most suitable form of grease-kettle is one that has a diameter at the bottom, of two-thirds the diameter at the top and a depth equal to the top diameter. The bottom and the side (or rather periphery), to a distance of one-quarter of the total height, should be jacketted so as to withstand a working steam-pressure of, say, fifty pounds per square inch. The inside of the kettle must be fitted with a perpendicular shaft fitted with arms or vanes (so that the lowest ones just clear the bottom, while another pair rotate at about the centre of the perpendicular height, and another pair about a foot from the top) and surmounted by bevel gears or other mechanical device for transmitting a rotary motion. The arms should be made of either cast or wrought iron and be about five times as wide as they are thick, and should be so attached to the shaft that as they rotate they depress the contents of the kettle; and the rate of rotation should not exceed thirty revolutions per minute. (See section of grease kettle, Plate II.)

In the manufacture of cup greases it is of the greatest importance that the milk of lime should be made from freshly slaked lime—lime bought ready slaked is of no use. Nothing but the best quality (preferably Jointa) lump lime should be used, and this is to be placed in an iron tank (provided with an open-end steam pipe), together with the requisite quantity of water, about an hour before the milk of lime is required. The mixture of lime and water should be boiled by the aid of steam, and after the slaking is completed the contents of the tank must be immediately (without allowing any settling) passed through a 40-mesh sieve in order to separate any grit, etc.

Supposing the capacity of the kettle to be 2,000 gallons of finished grease : 2,000 pounds of horse fat having been introduced and the stirrer started, steam is turned into the jacket, and when the temperature of the fat reaches about 180° F. the milk of lime (made by slaking two hundred pounds of lime in one hundred and forty gallons of water) is introduced and, the loose-fitting covers of the kettle being in place, the heat is gradually increased to the boiling-point. The contents of the kettle are now to be kept gently boiling and constantly stirred, and after about two hours a stick should be dipped in, then withdrawn and allowed to cool for a few seconds. If on examination it is found that soap is adhering to the stick, the kettle must be closely watched and its contents sampled every ten or fifteen minutes by means of the stick. The samples of soap thus obtained must be very carefully examined, and as soon as they cease to show any trace of free oil and are dull in appearance, it is a sign that the saponification is completed, and from now on the soap must be sampled every few minutes and a little of the soap must be rubbed down the palm of one hand by the fingers of the other, as soon as the soap yields only a few minute drops of water by this treatment, it is time to shut off the steam from the jacket and commence the addition of the mineral oil. A light-coloured mineral oil of 885 to 910 specific gravity is generally used, and about one hundred gallons that are at a temperature of about 190° F. should be slowly (taking about half an hour) sprayed on to the contents of the kettle; after this oil has been thoroughly stirred into and amalgamated with the soap, the further reduction or softening of the soap is accomplished by slowly spraying cold mineral oil on to the mixture and continuing to do so until the required consistency of grease is obtained, after which the stirring may be advantageously continued for about an hour.

Many failures in grease-making and much of the lumpy grease is attributable to the entirely wrong practice often adopted of starting to soften the lime-soap by the addition of cold instead of hot mineral oil. The fact that the addition of a cold oil to a high melting-point hot soap must result in the congealing of some of the soap and necessitate a high heat to remelt it, seems so obvious that it is astonishing so many grease-makers work along, year after year, and persistently keep in the old rut that leads to bad results and expensive manufacture. After the soap has been once softened by the addition of hot oil it is equally obvious that it would be unwise to continue the addition of hot oil, because the soap is sufficiently reduced in melting-point to preclude the likelihood of the mixture congealing if cold oil is gradually added, and, as it is desirable to cool the grease and have it ready for the market as soon as possible, the addition of cold oil at an early stage is decidedly advantageous. Also, a number of experiments made by the author demonstrated the fact that if, after the soap has been softened with hot oil, cold oil is used, a larger quantity of oil can ultimately be added than if hot oil is used from start to finish.

If it is found that when the soap is forming (and before any mineral oil has been added) that it commences to get "bright" while there is yet "free" or unsaponified fatty oil present, it is a sign either that the water has been boiled off too quickly or that sufficient lime has not been used. The first thing to be done in a case of this sort is to add water in small quantities at a time, and if, after boiling and stirring for an hour or more, there are no signs of any improvement, milk of lime must be added in small quantities until all trace of free oil has disappeared and the soap has assumed a dull appearance.

As previously stated, the soap must contain more moisture if a soft grease is required than if a hard one is to be made. If it is found that too much moisture has been driven from the soap to enable the grade of grease required being made, it is wiser to stop further addition of mineral oil, and barrel the thicker grease and hold it in stock rather than attempt to go on thinning it to the grade in view. If, however, so much mineral oil has been inadvertently added as to cause a "separation," it is often possible to reunite the mineral oil and soap by a very cautious addition of small quantities of hot water, together with, in extreme cases, the aid of gentle boiling. Should such an excess of mineral oil have been added that the reunion cannot be brought about in this way, the only alternative is to run off the contents of the kettle and work it off gradually into fresh batches of fat and lime. Even though the reunion can be effected, it must be remembered that the resulting grease is very much softer than would be obtained had the addition of water and additional boiling not been necessary, and, therefore, the yield of a given grade of grease will be diminished.

When it is desired to make a boiled grease from cotton or rape-seed oils, it will be necessary to use about twice as much lime as in the case of horse fat, and it will also be found that the saponification takes place very much more slowly. On the other hand, however, both cotton and rape-seed oil soaps yield a much larger percentage of grease than a horse-fat soap does, and are usually more pleasing to the eye and certainly more pleasing as regards smell.

It is a mistake to run off the grease from the kettle too hot, and much is to be gained both in appearances and "body" by allowing it to stand for twelve hours or so before removing it from the kettle.

Some grease-makers run the grease as soon as made into long semicircular, open-top tanks and paddle it for a number of hours until cold, and then pass it between steel rolls (to reduce any lumps) and pack it. This method of working is antiquated and very unsatisfactory and expensive. In the first place, running the grease boiling hot from the kettle and paddling it in an open vessel fills it with air bubbles, and, consequently, the paddling has to be continued until all the air has been squeezed out again—otherwise the grease will be dull.

Secondly, the above procedure results in the formaion of hard lumps due to some of the grease being cooled by contact with the iron tank and not being detached and mixed with the rest by the paddles. These hard lumps cannot be properly reduced or softened by passage between the rollers, and have to be worked over again with a fresh batch of raw material. Besides the above objections to this method of working there is another and more serious one, namely, the labour required is far greater than when the following method of handling the grease is adopted :—

The draw-off from the kettle should be in the centre of the bottom, and to it should be connected (by as short a piece of pipe as possible) a rotary pump which must discharge through a straight or curved (on no account use elbows or sharp turns) pipe into a cooler fitted internally with a spiral conveyer or "creeper," and from here the grease must be passed through a paint or putty mill, after which it is ready for packing.

The cooler may be cheaply and efficiently constructed by taking a piece of 6-inch wrought-iron pipe about 12 feet long (threaded at both ends), and a pair of 6-inch flanges having been placed in a lathe and recessed to a depth of half an inch, so that the ends of an 8-inch pipe will fit into the recesses, and a length of 8-inch pipe having been slipped over the 6-inch, and the recesses in the flanges having been fitted with thin rubber rings, the flanges are screwed on to the 6-inch pipe until the rubber in the recesses makes a tight joint between the 8-inch pipe and the flanges, and thus forms a water-tight annular space between the two pipes. All that remains to be done now is to tap a ½-inch (pipe) hole at the bottom of the outlet end and top of the inlet end of the 8-inch pipe for, respectively, the inlet and overflow of the cooling water; and to provide the 6-inch pipe with a connection for the ingress of the grease at one end and a connection at the other end that forms an exit for the grease into the mill and a stuffing-box for the conveyer shaft.

The best form of mill is one provided with stone rather than iron grinding surfaces, and it should be constructed to run at about three hundred revolutions per minute, and without undue friction of its bearings when working under a pressure of 100 pounds per square inch.

A mill of 18 inches diameter should put through (under about 70 pounds pressure per square inch) about 10 barrels of thick and 20 to 25 barrels of thin grease per hour.

Two men, with the above plant, can turn out and barrel ready for the market an average of 25 barrels of grease a day every day in the week, and in order to do the same work by the old method at least eight men would be required, besides three pairs of steel rollers each 10 feet long, and two cooling cylinders of 2 ft. 6 in. radius and 10 to 12 feet long, fitted with mechanical paddles. Also, the former method is clean and tidy, while the latter is the very reverse, even when the best of care is taken.

Cup greases are nearly always "perfumed," and generally by the addition of so-called oil of myrbane (*i.e.*, nitro-benzole). The perfuming should be done by sprinkling the myrbane over the grease as it is passing from the mill or rolls into the barrel or other package, and about one quart of perfume is sufficient for about forty-five gallons of grease. On no account must the myrbane be added to the grease in the kettle, because the temperature of the latter will be more than sufficient to volatilise the whole of the perfume long before the kettle can be emptied. The practice of perfuming greases undoubtedly originated from a desire on the part of the manufacturer to mask the origin of the grease, but such addition, for that purpose, must have been the result of ignorance, because it stands to reason that a volatile perfume only has to be subjected to a slight heat in order to entirely remove it, and, that having been done, a further increase in heat will soon tell the experienced nose whether the "base" of the grease is horse-fat, cotton-seed oil, etc., etc.

Engine Greases.

These greases differ entirely from "cup" greases in that they are usually made from "tallow" and lard oil and saponified by means of "caustic soda"; they are therefore soluble soaps. In colour they are white, or nearly so, while the consistency is very much harder than a cup grease.

The manufacture is accomplished by boiling, with constant agitation, a mixture of the fats or fatty oils with caustic soda solution of 1.075 specific gravity. The operation requires very close attention, and as soon as the soap hardens immediately on being placed on a moderately cool (50° to 65° F.) surface, the steam must be shut off from the jacket of the pan, and cold mineral oil quickly added with violent agitation. As soon as sufficient mineral oil has been added to reduce the soap to the required consistency, it is important that the contents of the kettle should be cooled as quickly as possible, and therefore, in order to secure the best results the bottom of the kettle should be connected to a cold-water supply, while a connection near the top carries off the water that is circulated through the jacket of the kettle until the whole of the constantly agitated grease is reduced to a temperature of about 90° F. When this temperature is reached the jacket is emptied, and then myrbane is usually added, and the grease, having been stirred sufficiently to ensure an equal distribution of the perfume, is ready for packing.

These greases often contain, besides tallow and lard oil:—

>Neatsfoot oil,
>Japan wax,
>Beeswax,

and other waxes and oils that the fancy may dictate.

Engine greases are undoubtedly good if scientifically made, but as this is seldom the case, they are more often harmful than otherwise. Usually they contain far too much soda, and in this case they absorb carbonic acid from the air and are then liable to eat into the bearings; if, on the other hand, they have not received sufficient soda, and therefore contain free fatty acids, they are equally as injurious.

The presence of free alkali in these greases may be easily detected by taking a piece about as big as a bean, and having dissolved it in about one-quarter ounce of boiling water, steep a piece of red litmus paper in the solution, and if it is turned blue it proves that free alkali is present. The intensity of the blue colour will depend on the excessiveness of alkali present; for instance, if the paper instantly turns dark blue it points to a large excess of alkali, but if it takes some minutes before a slight blue colouration is visible, it is proof that the grease is as nearly neutral as it is practicable to manufacture it.

Cup and engine greases include:

>Nos. 1 to 4 cup greases or "lubricants."
>Nos. 1 to 3 Albany greases.
>Sponge greases.
>Crankpin greases.

Gear greases.
Lubricating packing.
Plumbago and graphite greases, etc., etc.

Compounded Greases.

The manufacture of compounded greases requires but a few words, as it is essentially only a mechanical mixing of mineral, vegetable, or animal waxes, with mineral, vegetable, or animal oils.

A mixture of palm oil, tallow, and paraffin wax is often used for lubricating railway car-axles; while horse stearine mixed with mineral oil and paraffin wax is sometimes palmed off as a "cylinder" grease.

A compounded grease is an exceedingly poor article and an expensive one in the long run.

Greases cannot be tested for viscosity like oils can, and the determination of their body is essentially a matter of "feel." After a little experience in "working" a grease between the forefinger and thumb, the ability to detect differences in consistency is soon acquired, and this method is far more reliable than any tests that have so far been devised.

The manufacture of grease affords one of the very few exceptions to the rule, that the results obtained by working on the small scale seldom agree with those obtained by working on the large scale. The author has directed the manufacture of a very large number of batches (of from one to seventy barrels each) of the various greases mentioned in this work, and has also made exhaustive experiments on as small a scale as one pint, with the invariable result that the figures obtained from the pint batches tallied exactly with those obtained when the grease was made on the large scale. This is worth remembering, as one is often called upon to "match" a sample of grease and give a price per pound, and if it is known that the small-scale results will be similar to the large-scale ones, it enables the manufacturer to set a price on the grease as soon as his "matching" experiments are completed.

A few words of advice regarding the best methods of applying grease as a lubricant will doubtless be of value.

In the first place, the ordinary oil-cup is of no use when grease is to act as the lubricant, for the reason that the grease is too thick to pass through the small orifice connecting the cup and the bearing, unless it has pressure behind it; therefore a grease-cup must be constructed so that pressure can be brought to bear on the grease.

Some cups are made so that the cup proper is fixed to the bearing, and being threaded on the inside and fitted with a long threaded plug, the latter, when screwed down on to the grease contained in the cup, forces the grease into the bearings. Other cups are made so that the plug portion is fixed to the bearings and the cup portion is removable. This latter class of cups is undoubtedly the most suitable, for the reason that the cup part, being easily removed, can be cleanly filled and replaced while the machinery is in motion, and, if the cup portion is provided with a milled edge, it is at once both easier to handle and neater in appearance than the removable-plug cup.

Some grease cups are made so that the revolving shaft or spindle keeps a small pin (that passes upwards into the grease) in constant motion and thereby gradually and constantly works the grease down on to the spindle; this form of cup is objectionable for the reason that it causes the grease to become filled with air-bubbles and prevents its acting so efficiently as a lubricant as a grease that is undisturbed and passes into the bearings as a compact body.

Other cups are constructed with a snug-fitting piston which is kept constantly pressing on the grease and forcing the latter into the bearings by means of a spiral spring. Such a grease-cup is objectionable for the reason that, as it empties, the tension on the spring decreases, and therefore either too much grease is fed to the bearings when the cup is full, and the right quantity when the cup is partly empty, or the right quantity when full and too little after the cup is partially emptied.

Mr. Paul H. Grimm, of Glen Cove, Long Island, N. Y., is the inventor of a bearing that undoubtedly surpasses all others in economy of lubrication and reduction of wear and tear. It consists essentially of a hollow casting flanged internally at the ends, so that when a brass bush or sleeve is driven tightly home into the flanges it forms a water-tight annular space between the outside of the bush and inside of the casting. A hole is drilled and tapped through the centre of the top side of the casting and bush and the two are connected by means of a short piece of pipe, to the upper end of which the grease cup is attached, while the other end delivers the grease directly on to the spindle or shaft. Another hole is tapped for a ¼-inch pipe on the top of the casting near one end, while the bottom of the casting is tapped at the opposite end, thus providing for an inlet for a small stream of cold water which overflows through the top connection and can thus be kept constantly flowing around the bearing. With this form of bearing, one pound of grease will throughly lubricate the bearings of a 250 (16-candle) light dynamo for one week—running night and day from 5 P.M. Sunday to 5 P.M. Saturday. Machines that had been thus running for six months did not show the slightest trace of wear, and the wooden floor to which the dynamos were secured was as free from grease or oil as the day it was first laid, which is far more than can be said for the foundations of dynamos provided with ordinary bearings and lubricated with oil.

A heated bearing is a thing that often causes an engineer endless trouble and annoyance, besides resulting in a great waste of lubricant. After such bearings have been taken apart, trued, and thoroughly cleaned, further heating may often be prevented by mixing a small quantity of the best quality flower of sulphur with the lubricant, but if first-class flake graphite is at hand a mixture of about one-half ounce of it with a pound of the lubricant is preferable to the sulphur mixture.

The author has known gunpowder to be mixed with the lubricant with good effect in cases where neither sulphur nor graphite was at hand.

APPENDIX.

Some years ago the author wrote a short paper* on the results of his experiments with the action of oils on metals, and as these results have considerable bearing on the question of lubricants, this paper has been here reproduced, the author believing it will prove of value to those interested in the subject of lubricants.

Very little has yet been published in regard to the action of oils in common use on metals with which they are brought in contact when stored, transported, or employed for the lubrication of machinery; and as the subject is of importance, especially to manufacturers of compound lubricating oils, and to those who use such oils, the following record of the results of experiments extending over twelve months will, it is confidently hoped, be found of practical value.

Mr. C. W. Volney has made a few experiments on the action of oils on brass,† and finds that of olive, cotton-seed, and lard oils, the first has the most action and lard oil the least. Mr. W. H. Watson, in a paper on "The Action of Various Fatty Oils Upon Copper," ‡ points out that seal oil has more action than sperm oil.

* Journal of the Society of Chemical Industry, 1886, vol. v.
† Analyst, vol. viii., p. 68.
‡ Chemical News, vol. xxxvi., p. 200.

Dr. Stevenson Macadam has investigated the action of paraffin burning oils on metals,* and states that the action varied with different samples of such oils, and that the variation was not traceable to the presence of impurities. Engler, however, contends † that mineral oils have no action on metals if free from oxygen and air. In this connection it may be remarked that the action of hydrocarbons on metals has been generally considered to be due to impurities, such as water and slight traces of phenols and bases, which most commercial paraffin oils contain, on account of the difficulty of entirely removing them.

The experiments were made principally with a view to determine what fixed oils are best adapted for mixing with mineral oils for lubricating purposes.

Metals in ordinary use were employed, and, after being thoroughly cleaned, washed with ether, and dried, were weighed and placed in corked tubes together with 15 c. c. of the oil. These tubes were kept for twelve months at an average temperature of about 80° F. in the summer, and at 50° to 55° F. in the winter. It should be mentioned that the tallow oil was solid for four months during the winter of 1884, and one month at the commencement of the winter of 1885.

After the lapse of the twelve months the oils were all poured off from the metals and set aside for examination, and the metals, after being carefully washed with ether and cotton wool, were dried and weighed. On referring to the tabular statements of the results, it will

* Journal of the Chemical Society, vol. iii., p. 355.
† Chemical News, vol. xli., p. 284.

THE ACTION OF OILS ON VARIOUS METALS.

be seen that in the case of one iron and five zincs the metals gained in weight during the exposure to the action of the oils. This increase in weight is accounted for by the fact that, in the case of iron, the metal contained a flaw which had not been noticed and which became filled with oxide, and therefore could not be thoroughly cleaned; in the case of the zinc, block metal had been inadvertently used; and in all five cases the rough surface had become covered with a deposit which could not be removed.

Table IV. shows the weights of the different metals before and after exposure to the action of the oils, and the percentage loss sustained.

TABLE IV.

Showing the Action of Different Oils on a Particular Metal.

Name of Metal.	Weight in Grammes.		Percent-age Loss.	Name of Oil.	Remarks on Oils.	Remarks on Metals.
	At Start.	At Finish.				
Iron..	3.713	3.712	0.02693	Mineral Lub.	Slightly darkened in colour	
	3.200	3.1945	0.17187	Olive		
	3.395	3.3935	0.04418	Rape		
	3.3975	3.3905	0.20603	Tallow	Contained reddish precipitate	Partly rusted
	3.148	3.1445	0.11111	Lard		Bottom half covered with rust
	2.928	2.929	—?—	Cotton-seed	Much thickened and contained reddish-brown deposit	Showed traces of rust
	3.256	3.253	0.09210	Sperm		
	3.2575	3.2535	0.12279	Whale		Showed signs of rust
	3.248	3.2475	0.01539	Seal		
Brass	10.182	10.1795	0.02455	Mineral Lub.	Distinctly darkened in colour	
	11.230	Lost	—?—	Olive	Coloured bright emerald green	
	10.444	10.444	— —	Rape		
	8.114	8.110	0.04929	Tallow	Contained reddish-black precipitate	
	9.029	9.021	0.08860	Lard	Contained slight greenish-black sediment	
	8.9545	Tube broken,		Cotton-seed		
	7.2905	7.2885	0.02743	Sperm		
	9.814	9.8115	0.02547	Whale		
	10.514	10.513	0.00951	Seal	Surface gelatinised for ½-inch	Surface covered with dark film

TABLE IV.—*Continued.*

Name of Metal.	Weight in Grammes.		Percentage Loss.	Name of Oil.	Remarks on Oils.	Remarks on Metals.
	At Start.	At Finish.				
Tin	3.864	3.863	0.02587	Mineral Lub.		
	3.0535	3.0530	0.01637	Olive		
	2.884	2.884	— —	Rape		
	2.941	2.940	0.03400	Tallow	Reddish in colour	
	2.530	2.5255	0.01976	Lard		
	3.0385	3.0245	0.46075	Cotton-seed		
	3.512	3.511	0.02847	Sperm		
	3.4525	3.4525	— —	Whale		
	2.352	2.350	0.08508	Seal		
Lead ...	12.483	12.476	0.05607	Mineral Lub.	Distinctly darkened in colour	
	8.248	8.2445	0.04243	Olive	Had a greenish tinge	Thick slimy deposit on surface
	6.847	6.8335	0.19716	Rape		Thick slimy deposit on surface
	9.534	9.518	0.16782	Tallow		Slight slimy deposit on surface
	8.214	8.191	0.28000	Lard		Thick slimy deposit on surface
	7.7315	7.7260	0.07113	Cotton-seed		
	6.1965	6.1790	0.28241	Sperm		Thin yellowish-brown deposit on surface
	6.835	6.808	0.29502	Whale	Contained a flaky gelatinous deposit	Covered with a white deposit
	7.048	7.083	0.14198	Seal		
Zinc ...	0.499	0.499	—?—	Mineral Lub.		
	8.477	8.487	—?—	Olive	Contained a flaky gelatinous deposit	Covered with a white deposit
	6.855	6.872	—?—	Rape	Contained a flaky gelatinous deposit	Covered with a white deposit
	7.812	7.8245	—?—	Tallow	Reddish-brown in colour	Large deposit on surface
	7.251	7.2475	0.04826	Lard		
	7.952	7.989	0.16848	Cotton-seed		
	8.1575	8.0885	0.84585	Sperm		
	6.993	7.0215	—?—	Whale	Rather brown in colour	Surface covered with slime
	4.500	4.516	—?—	Seal		
Copper.	3.202	3.202	—?—	Mineral Lub.	Very much darkened in colour	
	1.251	1.2485	0.19984	Olive	Had a greenish colour, and very much thickened at surface	Covered with a thick blue deposit.
	1.183	1.180	0.25380	Rape		
	1.207	1.202	0.41425	Tallow		Thick deposit on surface
	1.052	1.0485	0.33269	Lard		Thick deposit on surface
	1.217	1.215	0.16433	Cotton-seed	Contained a yellowish deposit	
	1.193	1.192	0.06882	Sperm		
	1.190	1.1885	0.19605	Whale		
	1.204	1.200	0.33222	Seal		

THE ACTION OF OILS ON VARIOUS METALS. 47

In order to facilitate comparison of the results a diagram (Plate III.) has been prepared. It shows the percentage losses as given in Table IV. The results obtained with those metals which gained in weight in the manner described, and those which lost nothing, have been omitted.

On referring to Table IV. and Plate III. it will be seen that—

Iron is least affected by seal oil, and most by tallow oil.

Brass is not affected by rape oil, least by seal oil, and most by olive oil.

Tin is not affected by rape oil, least by olive oil, and most by cotton-seed oil.

Lead is least affected by olive oil, and most by whale oil; but whale, lard, and sperm oils all act to very nearly the same extent on lead.

Zinc seems, by the four actual weighings that were of any value, to be not acted on by mineral lubricating oil, least by lard oil, and most by sperm oil.

Copper is not affected by mineral lubricating oil, least by sperm oil, and most by tallow oil.

The table and diagram also show that—

Mineral Lubricating Oil has no action on zinc and copper, acts least on brass, and most on lead.

Olive Oil acts least on tin and most on copper.

Rape Oil has no action on brass and tin, acts least on iron, and most on copper.

Tallow Oil acts least on tin and most on copper.

Lard Oil acts least on zinc and most on copper.

Cotton-seed Oil acts least on lead and most on tin.

Sperm Oil acts least on brass and most on zinc.

Whale Oil has no action on tin, acts least on brass, and most on lead.

Seal Oil acts least on brass and most on copper.

From the foregoing results it will be seen that mineral lubricating oil has, on the whole, the least action on the metals experimented with, and sperm oil the most.

For lubricating the journals of heavy machinery, either rape or sperm oil is the best oil to use in admixture with mineral oil, as they have the least effect on brass and iron, which two metals generally constitute the bearing surfaces of an engine. Tallow oil should be used as little as possible, as it has considerable action on iron.

All the oils were examined for the respective metals with which they had been in contact, by the following methods:

Iron.—Agitate the oil with a dilute solution of nitric acid, draw off the solution, and evaporate to dryness on the water bath in a small porcelain basin. Take up with water, and add ammonia and sulphuretted hydrogen water.

Brass.—See Copper.

Tin.—Extract with dilute hydrochloric acid, evaporate solution to dryness, take up with water and a few drops of hydrochloric acid, and to this add a few drops of a mixture of a solution of ferric chloride and ferricyanide of potassium. If tin be present, a precipitate of "Prussian blue" will be given.

Lead.—Extract with dilute nitric acid and evaporate solution to dryness, dissolve in water, and test with sulphuretted hydrogen water.

Zinc.—Extract with dilute hydrochloric acid, evaporate, take up with water, and add ammonia and ammonium sulphide.

Copper.—Extract with dilute nitric acid, and evaporate (after the addition of a few drops of pure sulphuric acid) to a small bulk in a porcelain basin in order to remove all nitric acid. Dilute with a little water and place a new clean needle in the solution. The needle will soon become coated with a film of metallic copper if any be present.

The chemical examination of the oils appears to afford the most trustworthy guide in determining what metal is best adapted for the construction of storage tanks for the different oils; as in some cases where the percentage loss of metal arising from the formation of a deposit not taken up by the oil was high, only a trace (and in some cases no trace) of the metal was found in the oil.

Some of the oils have both a dissolving and depositing effect, while others have only one or the other. Slight traces of copper and zinc were found in the mineral lubricating oil, although the weighings showed the metals to have lost nothing; this indicates the importance of examining the oils for the metals.

INDEX.

ACTION of oils on metals, 4, 5, 7, 43
Albany greases, 37
Aluminium soaps, 10, 21
 tests for, 17
American oils, 1, 2, 16
Anti-friction metals, 10
Atomic weights of elements, 13
Axle greases—see set greases

BEARINGS, Grimm's patent, 40
Beeswax, 16, 37
Belt grease, 23
Black grease, 27
Boiled grease, 11, 23
 manufacture of, 28, 36
  old method, 31, 33
 theory of manufacture of, 11
Brass, action of oils on, 47

CAKE or cold-roll grease, 27
Castor-oil, 9, 16
Choice of lubricants, 2, 5, 8
Cold test, effects of, in viscosity of oils, 3
Color of oils, 16
Colza oil, 9, 16
Compounded greases, 11, 23, 38
 oils, 9, 20
Concentrated oils, 19
 superiority of, 20
Copper, action of oils on, 47
 testing for in oils, 49
Cordage oils, 21
Cotton-seed oil, 9, 16, 47
Crank-pin grease, 37

Cup greases, 11, 23
 manufacture of, 28, 37
  improved method, 33
 manufacture, labor required, 35
 manufacture, old method, 31, 33
 manufacture, theory of, 12
Curve grease, 27
Cylinder grease, 38
 oils (inferior), 21
 oils (pure), 16

DEBLOOMED oils, 5, 22
 tests for, 18
Decomposition of fatty acids, 7, 12
Degras, 16
Distillates, 19
Dynamo oil, 21

EFFECT of cold test on viscosity, 3
Electric-light oil, 21
Engine-grease, 28, 36
 characteristics of, 37
Engine oil, 21

FATTY acids, manufacture of, 7, 12
 effect of, on metals, 4, 5, 7
Fatty oils, 6
 characteristics of, 4, 6
 decomposition of, 7, 12
 tests for presence of, 15, 17
Ferry-slip grease, 27

GEAR grease, 27, 38
Glycerides, 6

INDEX.

Graphite, 10
  grease, 27, 38
Grease—Albany, 37
  belt, 23
  black, 27
  boiled, 11, 23, 28, 36
  cake, or cold-roll, 27
  compounded, 11, 23, 38
  crank-pin, 37
  cup, 11, 23, 28, 31, 33, 37
  curve, 27
  cylinder, 38
  engine, 28, 36, 37
  ferry-slip, 27
  gear, 27, 38
  graphite, 27, 38
  harness, 23
  hub-axle, 27
  leather, 23
  lubricating packing, 38
  mica, 27
  perfumed, 35
  pinion, 27
  plumbago, 38
  polishing, 23
  set, 11, 24, 27
  sponge, 37
  cooler, 34
  cups, 39
  how to avoid lumpy, 27, 31, 33
  kettle, 29, 38
  manufacture of, 23–38
  manufacture of, on large versus small scale, 38
  mill, 34
Greases, method of testing, 37–39
  why perfumed, 35
Grimm's patent bearing, 40
Gunpowder as a cooling medium, 41

HARNESS grease, 23
  oils, 21
Heated bearing, how to cool, 5, 41
Horse fat, 12, 16
Hub-axle grease, 27
Hydrocarbons, 1

INSOLUBLE soaps, 13
Iron, action of oils on, 4, 5, 7, 47
  tests for, in oil, 48

JAPAN wax, 16, 37

LARD oil, 9, 16, 47
Lead, action of oils on, 47
  soaps, 21
  tests for, in oil, 48
Leather grease, 23
  oils, 21
Lime part, 24
  slaking, 24
  soaps, 12
Linseed oil, 9
Lubricants, 37
  bad mineral, 10
  choice of, 2, 5, 8
  division of, 1
  effect of heat on, 5
  good mineral, 10
  test of fitness, 18
  where color is of no account in, 14
Lubricating packing, 38

MARSH gas, 1
Menhaden fish oil, 9, 16
Methods of using, 39
Mica grease, 27
Mineral lubricants, 10
  manufacture of, 19
Mineral lubricating oil, color no criterion of purity of, 13
  treated versus distilled, 13
Mineral oils, 1, 16
  characteristics of, 2
  tests for presence of, 18
  tests that influence choice of, 2
  tests of, 2, 15
Myrbane, oil of, 35

NEATSFOOT oil, 9, 16, 37

OIL, American, 1, 2, 16
  action of, on metals, 4, 5, 7, 43

INDEX. 53

Oil, castor, 9, 16
  color of, 16
  colza, 9, 16
  compounded, 9, 20
  concentrated, 19
  consistency of, at 75° F., 16
  cotton-seed, 9, 16
  cordage, 21
  cylinder, 16, 21
  debloomed, 5, 18, 22
  distillate, 19
  dynamo, 21
  electric light, 21
  engine, 21
  fatty, 4, 6, 12, 15, 17
  harness, 21
  lard, 9, 16, 47
  leather, 21
  linseed, 37
  melting point of, 16
  Menhaden fish, 9, 16
  mineral, 13, 15, 16, 18, 19
  myrbane, 35
  neatsfoot, 9, 16, 37
  olive, 9, 16, 47
  origin of, 16
  pale, 16
  palm, 9, 16
  polishing, 21
  rape-seed, 9, 16, 21, 47
  reduced, 19, 22
  rope, 21
  rosin, 11, 16
  Russian, 1, 16
  screw cutting, 8, 21
  shale, 16
  specific gravity of, 16
  sperm, 9, 16, 48
  seal, 48
  spindle, 2, 21
  tallow, 9, 16, 47
  type of, 16
  valve, 16
  viscosity of, 16
  West Virginia, 16
  whale, 10, 16, 48
  wool, 21

Oils, distinguishing between different fatty, 15
  table of origin, tests, etc., of, 16
  tests for metals in, 48
  tests of, 15
  that must not be used as lubricants, 9
  viscosity of, 2, 16
Olefiant gas, 1
Olefines, 1
Oleic acid, 6, 16
  glyceride, 6
Oleine, 12, 16
Olive oil, 9, 16, 47

PALE oils, 16
Palmatine, 12, 16
Palmitic acid, 6, 16
  glyceride, 6
Palm oil, 9, 16
Paraffines, 1
Perfumed grease, 35
Pine tar, 16
Pinion grease, 27
Plumbago, 10
  grease, 38
Polishing grease, 23
  oils, 21

RAPE-SEED oil, 9, 16, 47
Reduced oil, 19
    superiority of, 22
Rope oil, 21
Rosin oil, 16
Russian oils, 1, 16

SCREW cutting oils, 8, 21
"Set" measuring apparatus, 26
Set grease, 11, 23, 24
    effect of vibration on, 27
    effect of water on, 27
Shale oils, 1, 16
Soap, stearate-palmitic-oleate of lime, 12
Soaps, insoluble and soluble, 13
Soapstone, 10

Sperm oil, 9,
Spindle oils,
Sponge grea
Steam cylin(
    on, 7
    lubricat
Stearate-pa'
    12
Stearic aci(
    glyceride
Stearine, 1
Sulphur a:

TALLOW,
    oil, 9, 1
Testing π
    oils for

Sperm o
Spindle
Sponge
Steam c
    o
    lub
Stearat
    1
Stearic
    glyce
Stearin
Sulphu

TALLO
    oil,
Testin
    oils

WORKSHOP RECEIPTS,

FOR THE USE OF

MANUFACTURERS, MECHANICS,

AND

SCIENTIFIC AMATEURS.

FIRST, SECOND, THIRD, AND FOURTH SERIES,
EACH CONTAINING ABOUT 450 PAGES, WITH ILLUSTRATIONS, CROWN OCTAVO, CLOTH.

PRICE $2.00 EACH.

PRESS NOTICES.

This series comprises a store of information that is of service in every-day life. *National Druggist.*

We heartily recommend this cyclopædia of technical information to all of our readers, but especially to manufacturers and mechanics. *Building.*

The topics treated are very many, and minute details are given as to manipulation, and manufacture of materials that cannot be easily obtained in stores. *Builder and Woodworker.*

We cannot enumerate the heads of chapters, but it is safe to say to the reader that he will find in this series, reliable information upon every process known to the Arts in use at the present day. *Mechanical Engineer.*

Now ready. Containing **702** *pages, 8vo, cloth, with* **1420** *illustrations.*

SPONS'
MECHANICS' OWN BOOK.

A MANUAL FOR
HANDICRAFTSMEN AND AMATEURS.

The title of this work almost suffices to indicate the character of the contents. The various mechanical trades that deal with the conversion of wood, metals, and stone into useful objects are explained from an every day practical view.

The method of treatment of each branch is scientific yet simple. First in order comes the raw material worked upon, its characters, variations, and suitability. Then the tools used in working up the material are examined as to the principles on which their shape and manipulation are based, including the means adopted for keeping them in order, by grinding, tempering, filing, setting, handling, and cleaning. A third section, where necessary, is devoted to explaining and illustrating typical examples of the work to be executed in the particular material under notice. Thus the book forms a complete guide to all the ordinary mechanical operations; and whilst professional workmen will find in it many suggestions as to the direction in which improvements should be aimed at, amateur readers will be glad to avail themselves of the simple directions and ingenious devices by which they can in a great degree overcome the disadvantage of a lack of manipulative skill.

Price **$2.50** *postpaid.*

26 PLATES, 5¼ in. x 8 in. PRICE 50 CENTS.

SYSTEM
OF
EASY LETTERING.

BY

J. HOWARD CROMWELL.

AUTHOR OF "A TREATISE ON TOOTHED GEARING," "A TREATISE ON BELTS AND PULLEYS."

This useful work gives twenty-six different forms of alphabets, all constructed on the same general system. The space to be lettered is to be divided into parallelograms or squares, as the case may be, and within these as a guide the different letters are drawn and inked. The guide squares, which have been made in pencil, are then erased, leaving the final letters. On so simple a basis as this quite a variety of effects are produced in flat and block letters. *Scientific American.*

This little book will be appreciated by draftsmen who wish to use plain letters (and yet somewhat different from the ordinary run of letters) for the titles on drawings. The book will also be valuable and useful to any one who has had no practice in lettering, as the easy method given for forming the letters will enable a person to form the letters correctly, and with a little practice to do so quickly. *American Machinist.*

Mr. Cromwell has done a good work in getting out this little book, providing a system of lettering which is both neat and easy of execution. Instead of giving the conventional series of ornamental letters, many of which take more time than the drawing is worth, we have here several alphabets which give sufficient variety for all ordinary purposes, and yet are so simple as to be within the range of ability of the freshest apprentice. We recommend the book to every draftsman. *Mechanics.*

After forty years experience in mechanical and architectural drawing in this and other countries, we consider this among the best, most practical and useful books that treat on lettering for ordinary practical purposes. *Master Steam Fitter.*

Designs are given in great variety, and the lines of projection given with them enables any one to enlarge or reduce with great ease and exactness. *Manufacturers Gazette.*

THE BEST AND CHEAPEST IN THE MARKET.

ALGEBRA SELF-TAUGHT.

BY

W. PAGET HIGGS, M.A., D.Sc.

FOURTH EDITION.

CONTENTS.

Symbols and the signs of operation. The equation and the unknown quantity. Positive and negative quantities. Multiplication, involution, exponents, negative exponents, roots, and the use of exponents as logarithms. Logarithms. Tables of logarithms and proportional parts. Transportation of systems of logarithms. Common uses of common logarithms. Compound multiplication and the binomial theorem. Division, fractions and ratio. Rules for division. Rules for fractions. Continued proportion, the series and the summation of the series. Examples. Geometrical means. Limit of series. Equations. Appendix. Index. 104 pages, 12mo, cloth, **60c.**

See also **Algebraic Signs**, Spons' Dictionary of Engineering No. 2. 40 cts.

See also **Calculus**, Supplement to Spons' Dictionary No. 5. 75 cts.

Barlow's Tables of squares, cubes, square roots, cube roots, reciprocals of all numbers up to 10,000. A thoroughly reliable work of 200 pages, 12mo, cloth, $2.50.

Logarithms.—Tables of logarithms of the natural numbers from 1 to 108,000 with constants. By CHARLES BABBAGE, M.A. 220 pages, 8vo, cloth, $3.00.

Logarithms.—A. B. C. Five figure logarithms for general use. By C. J. WOODWARD, B.Sc. 143 pages, complete thumb index, 12mo, limp leather, $1.60.

Books mailed post-paid to any address on receipt of price

PRACTICAL PAMPHLETS.

Hints to Young Engineers Upon Entering Their Profession. By Joseph W. Wilson. 8vo, paper, 20 cts.

Designing Belt Gearing. By E. J. Cowling Welch. 16mo, paper, 20 cts.

Tables of the Principal Speeds Occurring in Mechanical Engineering expressed in metres, in a second. By P. Keerayeff. Translated by Sergius Kern, M.E. 16mo, paper, 20 cts.

The French Polishers' Manual. By a French polisher. 16mo, paper, 20 cts.

Cleaning and Scouring. A Manual for Dyers and Laundresses, and for Domestic Use. By S. Christopher. 16mo, paper, 20 cts.

The Cooking Range. Why is it my cooking range does not work properly, and why so extravagant with fuel? By F. Dye. 16mo, paper, 20 cts.

The Gas Consumers' Handybook. By William Richards, C.E. 16mo, paper, 20 cts.

Turner and Fitters' Pocket Book. For calculating the change wheels for screws on a turning lathe and for a wheel cutting machine. By J. La Nicca. 16mo, paper, 20 cts.

Screw-Cutting Tables, for the use of Mechanical Engineers, showing the proper arrangement of wheels for cutting the threads of screws of any required pitch, with a table for making the Universal Gas Pipe Threads and Taps. By W. A. Martin. Paper, 20 cts.

Ornamental Penmanship. A Pocketbook of Alphabets for Engravers, Sign Writers and Stone Cutters, containing 32 pages of alphabets. Paper, 20 cts.

The Handy Sketching Book, for the use of Draftsmen and Engineers, containing cross-section paper to EXACT eighth of an inch, with useful tables. Bound in boards, per dozen, $2.50. Each, 25cts.

The Kitchen Boiler and Water Pipes: Their arrangement and management—more especially their treatment during frost and how to avoid explosions. By H. Grimshaw. Second edition, illustrated, 8vo, paper, 40 cts.

PRACTICAL HANDBOOKS.

Engine Cylinder Diagram.
With (Meyer) cut-off at ⅙, ¼, ⅜ and ½ stroke of piston, with movable valves. By H. W. Weightman. On card, 25 cts.

How to Run Engines and Boilers.
A Handbook of Practical Instruction for Young Engineers and those ir charge of Steam Engines. By E. P. Watson. Illustrated, cloth, 1.00

The Fireman's Guide.
A Handbook on the Management of Boilers. Written in plain language for Young Firemen and those in charge of boilers. By K. P. Dahlstrom, M. E. Cloth, 50 cts.

The Corliss Engine and Its Management.
The best and most practical book on the Corliss Engine. By John T. Henthorn and C. D. Thurber. Plainly written, thoroughly practical, and fully illustrated. Cloth, 1.00

Ammonia Refrigeration.
Theoretical and Practical Ammonia Refrigeration. By I. I. Redwood, M. Am. Inst. M.E. A practical handbook for the use of those in charge of Refrigerating Machinery and Ice Plants. Fully illustrated and about 24 pages of valuable tables.. 12mo, cloth, 1.00

Manual of Instruction in Hard Soldering.
New edition, with an appendix on the Repair of Bicycle Frames, Soldering Alloys, Fluxes, and a chapter on Soft Soldering. By Harvey Rowell. Illustrated, cloth, 75 cts.

Practical Electrics.
A Handbook on the Construction, Fitting-up and Use of Telephones, Bells, Receivers, Batteries, Alarms, Annunciators, Dynamos, Motors, Carbons, Microphones, Phonographs, Photophones, Coils, and all kinds of electrical appliances. Fully illustrated, cloth, 75 cts.

The Magneto Telephone.
Its Construction, Fitting-up and Adaptability to Every-day Use. By Norman Hughes. Specially written in plain language for the use of beginners. Illustrated, cloth, 1.00

Syrups, Extracts, Flavorings, Phosphates, &c.
The Non Plus Ultra Soda Fountain Requisites. By G. Dubelle. A Handbook of Practical Receipts. Nearly 500 recipes. Cloth, 2.50

Algebra Self Taught.
The Best and Cheapest Algebra in the Market. By W. P. Higgs. Cloth, 60 cts.

A System of Easy Lettering.
By J. H. Cromwell. Containing 25 pages of alphabets. Well printed on good paper, oblong. Paper, 50 cts.

Pocket-books for Reference.

Molesworth and Hurst.—The pocket-book of pocket-books. Being Hurst's and Molesworth's pocket-book bound together in full Russia leather, 32mo., round corners, gilt edges, $5.00.

Cutler and Edge.—Tables for setting out curves from 100 feet to 5,000 feet radius. Useful for setting out roads, sewers, walls, fences and general engineering work, 16mo., cloth, $1.00.

Maycock.—Practical electrical notes and definitions, for the use of engineering students and practical men, 286 pages, illustrated, 32mo., cloth, 75c.

Thompson's Electrical Tables.—A valuable little reference book for engineers, electricians, motor inspectors and others interested in electrical engineering. Illustrated, vest pocket edition, 64mo., roan, 50c.

Dearlove.—Tables to find the working speed of cables; comprising also data as to diameter, capacity and copper resistance of all cores. 32mo., cloth, 80c.

Molesworth.—Metrical tables, weights and measures. 57 pages, 32mo., cloth, 60c.

Brooks.—French measures and english equivalents. 32mo., 40c.

Molesworth.—Pocket-book of useful formulæ and memoranda for civil and mechanical engineers. 781 pages, illustrated, 32mo, leather, $2.00.

George.—Pocket-book of calculation in stresses, etc., for engineers, architects and general use. 140 pages, illustrated, 32mo., cloth, $1.40.

Sexton.—Pocket-book for boiler makers and steam users, board of trade surveyors, and the general steam-using public. 287 pages, illustrated, 32mo., leather, $2.00.

Hurst.—A hand-book of formulæ, tables and memoranda for architectural surveyors. 477 pages, illustrated, 32mo., leather, $2.00.

Jordan.—Tabulated weights of angle, tee and bulb iron and steel and other information for the use of naval architects, shipbuilders and manufacturers. 579 pages, 32mo., leather, 1896, $3.00.

Spons'.—Tables and memoranda for engineers, by Hurst. The vest pocket edition, 64mo., roan, 50c.

Mackesy.—Tables of barometrical heights to 20,000 feet, specially adapted for the use of officers on service, civil engineers and travellers. With 3 diagrams, 32mo., cloth, $1.25.

Bayley.—Pocket-book for chemists, chemical manufacturers, metallurgists, dyers, distillers, brewers, sugar refiners, photographers, students, etc., etc. 32mo., roan, $2.00.   In the press.

Books mailed post-paid to any address in the world on receipt of price.

BJÖRLING'S BOOKS ON PUMPS.

The Construction of Pump Details. Windbores, foot-valves and strainers, clack-pieces, bucket-door-pieces and H-pieces, working barrels and plunger-cases. Plungers and rams. Piston and plunger. Bucket and plunger. Bucket and valves. Pump rods and spears. Spear-rod guides, etc. Valve-swords, spindles and draw-hooks. Set-offs. Pipes, pipe-joints and pipe stays. Pump slings, guide rods and guides. Kites, yokes and connecting-rods, L-bobs, T-bobs, angles or V-bobs, and balance-beams. Rock-arms and fend-off beams. Cisterns and tanks. Minor details. Index.
208 pages, 278 illustrations, 12mo., cloth. $3.00.

Practical Handbook on Pump Construction. Principles of the action of a pump. Description of pumps. Designing pumps. Materials for pumps. Designing pump valves Buckets, pistons, cup-leathers, air-vessels. Rules and formulas. Hydraulic memoranda. Tables. Index.
86 pages, 9 folding plates, 12mo., cloth. $1.50.*

Practical Handbook on Direct-acting Pumping Engines, and steam pump construction. Advantages and disadvantages of direct-acting steam pumps. Classification. Description of different types of valve gears. Rules and formulas. Index.
123 pages, 64 pages of tables, 20 illustrations, 12mo., cloth. $2.00.

Pumps, Historically, Theoretically and Practically Considered. Classification of pumps. Action of different types of pumps. Advantages and disadvantages of different types of pumps. General remarks on pumps. Remarks on designing pump valves. Index.
231 pages, 156 illustrations, 12mo., cloth. $2.50.

Water or Hydraulic Motors. Hydraulics relating to water motors. *Water Wheels.*—Breast, overshot, Pelton, general remarks on. *Turbines.*—Outward, inward, mixed, parallel and circumferential flow. Details of turbines. Regulation of turbines. *Water Pressure Engines.*—Reciprocating, rotative, oscillating, rotary, general remarks and rules for. *Hydraulic Rams.*—Without air-vessels in direct communication with the drive-pipe. With air-vessels, etc. Details of. Rules, formulas and tables for. Hydraulic pumping rams. Hydraulic ram engines. Measuring water in a stream and over a weir. Index.
287 pages, 206 illustrations and 12 tables, 12mo., cloth. $3.50.

See also **Pumps and Pump Motors.**

See also **Pumps,** Nos. 84 and 85, Spons' Dictionary, 40c. each, and **Pumps,** Suplement to Spons' Dictionary, No. 15, 75 cents.

See also **Hydraulics and Hydraulic Machinery,** Spons' Dictionary, Nos. 59, 60, 61, 62 and 63, 40c. each, and **Hydraulics,** Suplement to Spons' Dictionary, No. 12, 75 cents.

Books mailed post paid to any part of the world on receipt of price.

STRUCTURAL IRONWORK.

Structural Ironwork. The practical designing of structural ironwork. By Henry Adams, M. Inst. C.E. First series revised and enlarged. CONTENTS OF CHAPTERS. 1.—Small cast-iron girder. 2.—Flitched beam. 3.—Riveted joints. 4.—Wrought iron girder. 5.—Rolled joist. 6.—Trussed beam. 7.—Cast-iron bridge girder. 8.—Cast-iron stanchion 9.—Cast-iron column. 10.—Lattice girder bridge. 11.—Wrought iron roof truss. 12.—Specification tests and table of working stresses. 190 pages, folding plates, $3.50.*

Strains in Ironwork. A course of eight elementary lectures delivered before the Society of Engineers. By Henry Adams, M. Inst. C.E. 65 pages, 8 folding plates, 12mo, cloth, $1.75.

Stresses in Girder and Roof Trusses. For both dead and live loads by simple multiplication, with stress constants for 100 cases. For the use of civil and mechanical engineers, architects and draughtsmen. By F. R. Johnson, C.E. As in a work of this kind accuracy is the first consideration, no effort has been spared to avoid errors. The stress constants have been, as far as possible, determined in two or three different ways, and the results compared. The signs + and — have also been very carefully checked. 215 pages, 28 plates, 12mo, cloth, $2.50.

Plate Girder Railway Bridges. By Maurice Fitzmaurice. CONTENTS.—Strains in beams and girders. Strains in girders. Strains in solid beams. Loads on bridges. Working stresses in steel and iron. Materials. Kinds of bridges. Depth of girders. Ordinary sizes of plates and angle bars in steel and iron. Joints in plates, angles, etc. Three main girder bridge. Two main girder bridge. Conclusion. Index. 104 pages, 36 illustrations and 4 large folding plates, 8vo, cloth, $2.40.

Cylinder Bridge Piers. Notes on, and the well system of foundations. Especially written to assist those engaged in the construction of bridges, quays, docks, river-walls, weirs, etc. By John Newman, C.E. CONTENTS OF CHAPTERS. 1.—General design. 2.—To determine the required diameter of a cylinder bridge pier. 3.—Load on the base. 4.—Surface friction. 5.—Sinking cylinders—general notes. 6.—Sinking cylinders, staging, floating out. 7.—Removing obstructions in sinking and "righting" cylinders. 8.—Kentledge. 9.—Hearting. 10.—The compressed air method of sinking cylinders. 11.—Limiting depth, air supply and leakage. 12.—Effects of compressed air on men. 13.—Air locks. 14.—Working chamber and method of lighting it. 15.—Excavating and dredging apparatus for removing the earth from the interior of a cylinder or well. 16.—Notes on some dredging apparatus used in sinking bridge cylinder and wells. 17.—Sand-pumps, suction, compressed air and water-jet dredgers. 18.—The well system of foundations for bridge piers, abutments, quays and dock walls. Index. 8vo, $2.50.

Railway Bridges and Structures.

Long Span and Short Span Railway Bridges.—Comprising investigations of the comparative theoretical and practical advantages of the various adopted or proposed type systems of construction, with numerous formulas and tables giving the weight of iron or steel required in bridges from 300 feet to the limiting spans; to which are added similar investigations and tables relating to Short Span railway bridges. By B. Baker, Asso. Inst. C. E.   157 pages, 2 large folding plates, 12mo., cloth, $2.00

Strains on Braced Iron Arches and Arched Iron Bridges.—By A. S. Heaford. 39 pages of descriptive matter, diagrams and tables, and seven large folding plates. 8vo., cloth, $2.40

Triangulation and Measurements of the Forth Bridge.—By R. E. Middleton, M. Inst. C. E.   A description of the measurements of a base line, the triangulation of stations therefrom and the setting out of the foundations and portions of the steel work of the Forth Bridge by means of direct measurements with standard rods, indirect measurements with steel wire, and by triangulation ; accompanied by descriptive remarks on the instruments used, along with tables showing the instrumental and personal errors, and the degree of accuracy obtained in the several measurements and standards.   48 pages, with illustrations and tables, 8vo., cloth. $1.20

Iron and Timber Railway Superstructures, And general works ; giving dimensions and quantities for the standard 4 feet 8¼ inch gauge, and the metre 3 feet 3⅜ inch gauge ; also applications for light railways, steam tramways, etc.   With some earthwork tables, and outline of a specification and requirements.   By J. W. Grover, M. Inst. C. E., 47 pages, tables and diagrams, and 36 large lithographed scale drawings fully explained.   Folio, cloth, $17.00.

Railway Bridges, Estimates and diagrams of, for turnpike, public and occupation roads in the embankments of double or single lines, and cuttings ot double lines, with form for calculating quantities in skew structures, etc.   Also culverts of various dimensions.   By J. W. Grover, M. Inst. C. E.   Also diagrams of station buildings designed and executed by the late I. K Brunel and R. J. Ward.   Second edition, enlarged with 12 pages of tables and 47 large lithographed scale drawings, many of them colored.   Folio, cloth, $12.50

The New Tay Bridge.—A course of lectures delivered at the Royal School of Military engineering at Chatham.   By Crawford Barlow, M. I. C E.   *Contents :* A Short Sketch of Metal Bridge Designing.   Length of the Tay Bridge.   History of the Old Bridge.   General Description of the New Bridge.   Design of the Piers.   Design of the Girders.   Flooring, etc.   Provisions for Wind Pressure.   Permanent Way and Safety Guard.   Arrangements for Variation of Temperature.   Materials Used in Construction.   Description of Construction.   46 pages, tables and illustrations, one large folding plate to scale, and 22 large plates.   Folio, cloth, $8 50.

SCREW CUTTING, TURNING, &c.

Screw Cutting. Tables for Engineers and Mechanics, giving the values of the different trains of wheels required to produce Screws of any pitch. Calculated by Lord Lindsay. 8vo, oblong. 80c.

Screw Cutting. Turner's Hand-book on Screw Cutting. Coning, &c., &c, with Tables, Examples, Gauges, and Formulæ. By Walter Price. 8vo, cloth. 40c.

Screw Cutting. Screw Cutting Tables, for the use of Mechanical Engineers, showing the proper arrangement of Wheels for cutting the Threads of Screws of any required pitch, with a Table for making the Universal Gas-pipe Threads and Taps. By W. A. Martin. Sixth edition, oblong, cloth. 40c.

Screw Cutting. Tables showing the screws that can be cut on slide lathes. By H. L. Change wheels for screw cutting. Whitworth's threads. 31 pages, 8vo, boards. 40c.

Screw Cutting. Turners' and fitters' pocket-book for calculating the change wheels for screws on a turning lathe and for a wheel cutting machine. 16mo, sewn. 20c.

Turning. The Practice of Hand-Turning in Wood, Ivory, Shell, &c , with Instructions for Turning such work in Metal as may be required in the Practice of Turning in Wood, Ivory, &c., also an Appendix on Ornamental Turning. (A Book for Beginners.) By Francis Campin. Third edition. *Contents:*—On Lathes, Turning Tools, Turning Wood, Drilling, Screw-cutting, Miscellaneous Apparatus and Processes, Turning particular forms; Staining, Polishing, Spinning Metals, Materials, Ornamental Turning. 300 pages, 99 illustrations, 8vo, cloth. $2.00

Millwright's Guide. The Practical Millwright's and Engineer's Ready Reckoner; or Tables for finding the diameter and power of cog-wheels, diameter, weight, and power of shafts, diameter and strength of bolts, &c. By Thomas Dixon. Sixth edition. *Contents:*—Diameter and Power of Wheels; Diameter, Weight, and Power of Shafts; Multipliers for Steam used expansively; Diameters and Strength of Bolts; Size and Weight of Hexagonal Nuts; Speed of Governors for Steam Engines; Contents of Pumps; Working Barrels; Circumferences and Areas of Circles; Weight of Boiler Plates; French and English Weights and Measures. 93 pages, 12mo, cloth. $1.25

Mill Work. A Practical Treatise on Mill-gearing, Wheels, Shafts, Riggers, &c., for the use of Engineers. By Thomas Box. Third edition. *Contents:*—CHAP. 1. On Motive Power. 2. On Wheels. 3. On Shafts. 4. On Riggers or Pulleys. 5 On Keys for Wheels and Riggers. 6 Examples of Gearing in Practice. 120 pages, 11 plates, 12mo, cloth. $3.00

Water Supply, Plumbing, Drainage.

Hot Water Supply. A practical treatise upon the fitting of hot water apparatus for domestic and general purposes. By F. Dye. 82 pages, 25 illustrations, 12mo, cloth. $1 00.

Hot Water Fitting and Steam Cooking Apparatus. A guide for builders and others for the fitting and fixing of boilers and pipes for the circulation of hot water for heating buildings and for domestic purposes, with a chapter on boilers and fittings for steam cooking. By F. Dye. 92 pages, 23 illustrations, cloth, 50c.

Plumbing, Drainage and Water Supply, and Hot Water Fitting. By John Smeaton, C. E. Drainage. City wells. External plumbing. Internal plumbing and fittings. Tapping mains under pressure. Ornamental lead work. Heating. Hot water work, etc., etc. 236 pages, 217 illustrations, 8vo., cloth. $3.00.

Treatise on Modern Sanitary Appliances for Healthy Residences and Public Institutions. By F. Colyer, C.E. For the use of students, etc. General instructions. Drainage. Internal work. Water supply. Heating apparatus for buildings. Swimming baths. Sundry sanitary arrangements in the construction of residences, cleaning drains, etc. Electric light in residences and public institutions. Index, etc. 118 pages, 12mo., cloth, $2 00.*

Bad Drains and How to Test them. With notes on the ventilation of sewers, drains, and sanitary fittings, and the origin and transmission of zymotic diseases. By R. Harris Reeves. 68 pages, 3 plates, 12mo., cloth. $1.40.

Public Baths and Washhouses. By Robt. Owen Allsop. CONTENTS OF CHAPTERS:—1. Introduction. 2. Schedule of accomodation. 3. General arrangement of baths and washhouses. 4. Slipper baths. 5. Swimming baths. 6. The public washhouse. 7. The establishment laundry. 8. The engineer's department. 9. Water supply. 10. The heating of swimming baths. 11. The Turkish bath in public bathing establishments. 12. Baths for the poor. Index. 98 pages, 18 illustrations and 4 folding plates, 8vo, cloth, $2.50.

Hints to House Hunters and House Holders. By Ernest Turner. CHAPTER HEADINGS:—Househunters and Householders. 2. Requirements. 3. Situation and aspect. 4. Soil. 5. Construction. 6 Water supply. 7. Drains. 8. Ventilation and heating. 9. Warming and ventilation. 10. Light. 11. Dust. 12. Kitchen. 13. House hunting. 14. Lessors, lessees and their liabilities, etc. 161 pages, 26 illustrations. 12mo., cloth, $1.00.

Books mailed post-paid to any address on receipt of price.

PRACTICAL HANDBOOK
ON
GAS ENGINES.
With Instructions for Care and Working of the Same.

By G. LIECKFELD, C.E.

TRANSLATED WITH PERMISSION OF THE AUTHOR BY

Geo. Richmond, M.E.

TO WHICH HAS BEEN ADDED FULL DIRECTIONS FOR THE RUNNING OF

OIL ENGINES.

CONTENTS.

Choosing and installing a gas engine. The construction of good gas engines. Examination as to workmanship. As to running. As to economy. Reliability and durability of gas engines. Cost of installing a gas engine. Proper erection of a gas engine. Construction of the foundation. Arrangement for gas pipes. Rubber bag. Locking devices. Exhaust pipes. Air pipes. Setting up gas engines. Brakes and their use in ascertaining the power of gas engines. Theory of the brake. The Brauer band brake. Arrangement of a brake test. Explanation of the expressions "Brake Power" and "Indicated Power." Comparisons of the results of the brake test and the indicated test. Quantity of work consumed by external friction of the engine Distribution of heat in a gas engine. Attendance on gas engines. General remarks. Gas engine oil. Cylinder lubricators Rules as to starting and stopping a gas engine. The cleaning of a gas engine. General observations and specific examination for defects. Different kinds of defectives. The engine refuses to work. Non-starting of the engine. Too much pressure on the gas. Water in the exhaust pot. Difficulty in starting the engine. Clogged slide valve. Leaks in gas pipes. Unexpected stopping of engine. Irregular running. Loss of power. Weak gas mixtures. Late ignition. Cracks in air inlet. Back firing. Knocking and pounding inside of engine. Dangers and precautionary measure in handling gas engines. Examination of gas pipes. Precautions when :— Opening gas valves. Removing piston from cylinder. Examining with light openings of gas engines. Dangers in starting. Dangers in cleaning. Safeguards for fly-wheels. Danger of putting on belts. **Oil Engines.** Gas engines with producer gas. Gasoline and oil engines The "Hornsby-Akroyd" oil engine. Failure to start. Examination of engine in detail. Vaporizer valve box. Full detailed directions for the management of Oil Engines. Concluding remarks. 120 pages, illustrated, 12mo, cloth, $1.00.

MANUAL OF INSTRUCTION
—IN—
HARD SOLDERING,
—BY—
HARVEY ROWELL.

CONTENTS.

<u>Introduction. Utensils and Chemicals.</u>—The flame. Lamp. Charcoal. Mats. Blowpipes. Wash-bottle. Binding wire. Borax. Chemicals.

<u>Alloys for Hard Soldering.</u>—Spelter. Silver solder. Gold solder.

<u>Oxidation.</u>—Oxidation of metals. Fluxes. Anti-oxidizers.

<u>Structure of Flame.</u>—Oxidation of gases. The cone. Oxidizing flame. Reducing flame.

<u>Heat.</u>—Transmission. Conduction. Capacity of metals. Radiation. Application.

<u>The Process.</u>—The work table. The joint. Applying solder. Applying heat. The use of the blowpipe. Making a ferrule. Joints. To repair a spoon. Difficulties. To repair a watch case. Hard soldering with a forge or hearth. Hard soldering with tongs.

<u>Technical Notes.</u>—Preserving thin edges. Silversmiths' pickle. Restoring color to gold. Chromic acid. Steel springs to mend. Sweating metals together. Retaining work in position. Making joints. Applying heat. Preventing the loss of heat. Effect of sulphur, lead and zinc. To preserve precious stones. Annealing and hardening. Burnt iron. To hard solder after soft solder.

<u>Properties of Metals.</u>—Tables of specific gravity. Table of tenacity. Table of fusibility. Fusibility of alloys.

56 Pages, 12mo, Cloth, Price, 75 Cents.

CHARTS FOR

Low Pressure Steam Heating

FOR THE USE OF

ENGINEERS, ARCHITECTS, CONTRACTORS AND
STEAM FITTERS.

By J. H. KINEALY, M.E.
M. Am. Soc. M. E., M. Am. Soc. of H. and V. Eng'rs, &c., &c.

The author has long been in the habit of using charts to aid him in his work. Knowing the value of them in saving time, simplifying work and ensuring correct calculations he feels confident that they will be appreciated by engineers, architects and contractors, for whose benefit they have been compiled. Care has been taken to make the charts as clear and as easily understood and, above all, as accurate as possible. They have been based upon theoretical considerations, modified by what is considered to be good practice in this country.

CHART 1.—This chart is for determining the number of square feet of heating surface of a low pressure steam heating system, pressure not to exceed 5 lbs. per square inch by the gauge, necessary to supply the heat lost through the various kinds of wall surfaces of rooms. The chart is divided into four parts. CHART 2.—For determining the diameters of the supply and return pipes for a heating system CHART 3.—For finding the number of square feet of boiler heating surface and the number of square feet of grate surface for a boiler that is to supply steam to a steam heating system. CHART 4.—For determining the area of the cross section of a square flue, or the diameter of a round flue, leading from an indirect radiation heater to the register in a room to be heated.

Full details are given for the use of these cards.

These four charts are printed on heavy white card-board and bound together with cloth, size 13 in. by 9¼ in., **$1.00‡.**

These cards are securely packed for mail and sent to any part of the World on receipt of price.

MANUFACTURE OF GAS.

Gas Coal and Cannels, A treatise on the comparative commercial values of. By David A. Graham. *Contents:*—On ascertaining the relative commercial values of different qualities. The quantity of gas a ton can produce. Illuminating power. Rules for finding the makes and percentages. Weight, value, etc., of coke produced. Weight and strength of ammonical liquor. Tar. Illuminating power and production per ton, etc. Assigning values to the gas produced. Naphthaline. The cheapest coal from which standard gas can be made. On valuing gas coals and cannels. 100 pages, tables and 3 folding plates, 8vo., cloth. $3.00

Manufacture of Gas, A manual of the. from tar, oil and other liquid and hydrocarbons, and extracting oil from sewage sludge. By W. Burns, C.E. This little work is not intended to be a scientific disquisition on the destructive distillation of hydrocarbons. It is intended for the benefit of practical men who are interested in the subject. With two folding plates and four illustrations of the apparatus for making gas from coal tar. pages, 12mo, cloth. $1.50

Gas.—The gas fitters guide, with details of fittings. By John Eldridge. 12mo, sewed. 40c.

Gas.—Manual for gas engineering students. By D. Lee. 18mo, cloth. A practical little handbook. 40c.

Gas Manufacture.—A practical treatise on the manufacture and distribution of coal gas. By Wm. Richards, C.E. *Synopsis of Contents:*—Introduction; history of gas lighting; chemistry of gas manufacture, by Lewis Thompson; coal with analysis, by J. P. Lewis; hydraulic main; condensers; exhausters; washers and scrubbers; purifyers; purification; history of gas holder; tanks; brick and stone; concrete; cast-iron; compound annular wrought-iron; specifications; gas holders; station meter; governor; distribution; mains; gas mathematics or formulæ for the distribution of gas by Lewis Thompson; services; consumers meters; regulators; burners; fittings; photometers; carburization of gas; air gas and water gas; composition of coal gas, by Lewis Thompson; analysis of gas; influence of atmospheric pressure and temperature on gas; residual products; appendix; index. Illustrations, 4to, cloth. $12.00

Gas Works.—Their arrangement, construction, plant and machinery. By Frederick Colyer, M. Inst. C. E. *Contents of Chapters.*—1. Introduction. 2. Arrangement of a works. 3. Materials used. 4. Retort houses. 5. Engine and exhauster houses. 6. Retort house plant. 7. Coal machinery, hydraulic cranes and hoists. 8. Boilers, feed water apparatus. 9. Pumping machinery. 10. Purifying apparatus. 11. Travellers for purifier cover. 12. Gasholders. 13. Station meter house. 14. Photometer and testing rooms. 15. Mains, valves, etc. 16. Manufacture of sulphate of ammonia. 17. Useful memoranda. Index. 134 pages with numerous tables and 31 folding plates to scale. 8vo., cloth. $5.00

Gas. *See also* Spons' Dictionary of Engineering.

THEORETICAL AND PRACTICAL
Ammonia Refrigeration.

A Work of Reference for Engineers and others Employed in the Management of Ice and Refrigeration Machinery.

BY ILTYD I. REDWOOD,
Assoc. Mem. Am. Soc. of M. E.; Mem. Soc. Chem. Indus. Eng.

CONTENTS.

B. T. U. Mechanical Equivalent of a Unit of Heat. Specific Heat. Effect of Pressure on Specific Heat of Ammonia Gas. Specific Heat of Air with Constant Pressure. Specific Heat of Air with Constant Volume. Latent Heat. Latent Heat of Liquefaction. Latent Heat of Vaporization. Latent Heat of Water. Absolute pressure. Absolute Temperature. Absolute Zero. Effect of Pressure on Volume of Gases. Theory of Refrigeration. Freezing by Compressed Air. Freezing by Ammonia. Characteristics of Ammonia. Explosiveness. Tendency of the Gas to Rise. Solubility in Water. Action on Copper. 26° Ammonia. Anhydrous Ammonia. The Compressor. Stuffing-Boxes. Special Lubrication. Oil for Lubrication. Clearance Space, etc. Suction and Discharge Valves. Effect of Excessive Valve-Lift. Regulation of Valve-Lift. Separator, Condenser, Condenser-Worm, Receiver. Refrigerator or Brine Tank. Size of Pipe and Area of Cooling Surface. Expansion Valves. Working Details—Charging the Plant with Ammonia. Jacket-Water for Compressor. Jacket-Water for Separator. Quantity of Condensing Water Necessary. Loss due to Heating of Condensed Ammonia. Superheating Ammonia Gas. Cause of Variation in Excess Pressure. Use of Condensing Pressure in Determining Loss of Ammonia by Leakage. Cooling Directly by Ammonia. Brine. Freezing Point of Brine. Effect of Composition on Freezing Point. Effect of Strength on Freezing Point. Suitableness of the Brine. Making Brine. Specific Heat of Brine. Regulation of Brine Temperature. Indirect Effect of Condensing Water on Brine Temperature. Directions for Determining Refrigerating Efficiency. Equivalent of a Ton of Ice. Compressor Measurement of Ammonia Circulated. Loss of Well-Jacketed Compressors. Loss in Double Acting Compressors. Distribution of Mercury Wells. Examination of Working Parts. Indicator Diagrams. Ammonia Figures—Effectual Displacement Volume of Gas. Ammonia Circulated per Twenty-Four Hours. Refrigerating Efficiency. Brine Figures—Gallons Circulated. Pounds Circulated. Degrees Cooled. Total Degrees Extracted. Loss due to Heating of Ammonia Gas. Loss due to Heating of Liquid Ammonia. Calculation of the Maximum Capacity of a Machine. Preparation of Anhydrous Ammonia. Construction of Apparatus. Condenser-Worm. Why Still is Worked under Pressure. Best Test for Ammonia. Water from Separators. Lime for Dehydrator. Yield of Anhydrous from 26° Ammonia. Index.

150 Pages, 15 Illustrations and 24 Pages of Tables,
12mo, Cloth, $1.00.

SPON & CHAMBERLAIN,
Publishers of Engineering, Mining, Electrical and Industrial Books.
12 CORTLANDT ST., NEW YORK.

Lightning Source UK Ltd.
Milton Keynes UK
UKHW021250110722
405683UK00006B/1356